▼ ▼ ▼ ▼

HOLT

LIFE

SCIENCE

Teaching Resources

Unit 3
Chapters 8 and 9

HOLT, RINEHART AND WINSTON

Austin • *New York* • *Orlando* • *Chicago* • *Atlanta* • *San Francisco* • *Boston* • *Dallas* • *Toronto* • *London*

Some material in this work was previously published in HBJ LIFE SCIENCE, Teacher's ResourceBank, copyright © 1989 by Harcourt Brace & Company. All rights reserved.

Printed in the United States of America

ISBN 0-03-098385-1

2 3 4 5 6 7 8 9   085   96 95 94

Unit 3   Simple Living Things

Contents

CHAPTER 8 *Viruses and Monerans*

SECTION 1   Viruses

A. Directions: Write the correct term from the list for each description.

1. The organism invaded by a virus _____

2. The rapid spread of a disease through a large area _____

3. Provides the body with the ability to fight infection _____

4. Given to people to keep them from getting a disease _____

5. Carries the genetic, or hereditary, instructions _____

6. Billionths of a meter _____

7. A disease caused by a virus that can result in paralysis _____

8. A disease caused by a virus that weakens the human immune system _____

> AIDS
>
> immune system
>
> nucleic acid
>
> vaccine
>
> host
>
> nanometer
>
> epidemic
>
> polio

B. Directions: Circle the numbers of the phrases that describe characteristics of viruses.

1. Can reproduce only inside a living cell

2. Provide the body with the ability to fight infection

3. Considered hostile because they invade and then attack living cells

4. Responsible for the common cold and influenza

5. Are the material that carries hereditary instructions

6. Contain proteins and nucleic acids

7. Can prevent people from getting a disease

8. Are neither living nor nonliving

Name _____

Class _____ Date _____

Viruses and Monerans

SECTION 2 Monerans

A. Directions: Underline the term or phrase that makes the most sense.

1. Monerans do not have (chlorophyll, a nucleus, a cell wall).

2. (Vaccines, Monerans, Antibiotics) are chemical substances that can be used to kill or slow the growth of bacteria.

3. A chemical that gives color to the tissue of living organisms is called a (pigment, moneran, bacterium).

4. Bacteria can cause disease by destroying cells or giving off (viruses, toxins, nucleic acids).

5. (Protists, Bacteria, Antibiotics) are necessary for the decay of waste materials.

6. (Cheeses, Algae, Pigments) are produced with the help of bacteria.

7. Bacteria cannot survive unless they have (a plant or an animal as a host, a cold environment, enough food).

8. All bacteria can (make their own food, reproduce quickly, cause dangerous illnesses).

9. The cyanobacteria often found in bodies of water are incorrectly called "blue-green algae" because they are *not* (monerans, protists, always red or brown).

10. Cyanobacteria help to restore (oxygen, carbon, nitrogen) to the soil.

B. Directions: Circle the numbers of the phrases that describe characteristics of monerans.

1. Cause the common cold

2. Can live alone or in groups

3. Are much larger than viruses

4. Do not have a nucleus

5. Always give off toxins

6. Always have a cell wall

7. Are always killed by antibiotics

8. Are always blue or green

Name _____

Class _____ Date _____

CHAPTER
8   *Viruses and Monerans*

Chapter 8

Chapter Review

Directions: Circle the letter of the best term for each description.

1. A very small particle that can reproduce only inside a living cell
   a. moneran   **b.** bacteria   **c.** virus

2. Responsible for chicken pox, measles, and AIDS
   a. bacteria   **b.** viruses   **c.** epidemics

3. Found in the center of a virus
   a. protein   **b.** nucleic acid   **c.** chlorophyll

4. Covers the center of a virus
   a. protein   **b.** bacteria   **c.** cell wall

5. Spreads rapidly among the population in a large area
   a. vaccine   **b.** bacteria   **c.** epidemic

6. Can prevent people from getting diseases caused by a virus
   a. vaccine   **b.** antibiotics   **c.** immune system

7. Can cure a disease caused by bacteria
   a. vaccine   **b.** antibiotics   **c.** epidemic

8. Chlorophyll is one example of this
   a. cyanobacteria   **b.** pigment   **c.** bacteria

9. Have a cell wall but do not have a nucleus
   a. viruses   **b.** protists   **c.** monerans

10. What a virus needs to live and reproduce
    a. host   **b.** oxygen   **c.** another virus

11. One difference between viruses and bacteria is that bacteria can
    a. live alone   **b.** reproduce   **c.** be harmful

12. Cyanobacteria always have
    a. a blue-green color   **b.** a cell wall   **c.** harmful effects

HRW material copyrighted under notice appearing earlier in this work.

3

Name _____

Class _____ Date _____

Testing Disinfectants

(Textbook page 224) As you perform this investigation from your textbook, use this sheet to record your results and to answer the questions.

ANALYSES AND CONCLUSIONS

1. Which section(s) of the dish contained the variable(s)? The control(s)?

   _____

   _____

2. Which disinfectant appears to be the most effective? How can you tell?

   _____

   _____

3. The agar and the Petri dish were sterilized before this investigation. Why do you think this was necessary?

   _____

   _____

   _____

   _____

   _____

   _____

4. What is the purpose of a control in an investigation?

   _____

   _____

   _____

   _____

(continues)

INVESTIGATION 8 *Testing Disinfectants* (continued)

APPLICATION

Why do you think it is important to wash your hands thoroughly at the end of this investigation? List several ways in which this procedure applies to your everyday life.

_____

_____

_____

_____

_____

_____

_____

_____

Chapter 8

20 ▶ Shapes of Viruses

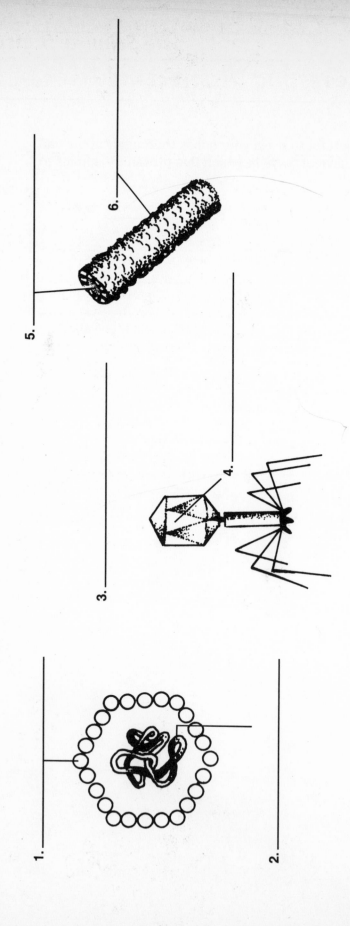

1. How are viruses similar to living things? How are viruses different from living things?

2. Describe the shapes of viruses.

Shapes of Viruses

Teaching Strategies

- Before introducing the transparency, ask the students to describe some of the shapes of living things in the world around them. After accepting their descriptions, remind the students that the organisms of the microscopic world display diversity in shape just as macroscopic living things do. Present the transparency and ask the students to recall what a virus is. (A virus is a nonliving material composed of genetic material and protein.) Ask the students to describe whether a virus is more likely to benefit or harm an organism it invades. (Viruses harm organisms by invading and destroying the cells of the organism.) Have the students describe the shapes of viruses shown as you point them out on the transparency. Ask them to identify the characteristics that distinguish one virus from another. (Shape, size, and the visibility of nucleic acid distinguish viruses.)

- After reminding the students that viruses are harmful to living things, distribute a worksheet to each student.

Questions

1. How are viruses similar to living things? How are viruses different from living things? (Viruses and living things are both composed of proteins and nucleic acids; viruses are not alive.)

2. Describe the shapes of viruses. (Viruses may have cylindrical, spherical, or spiral shapes as well as many other shapes.)

1. _____

    _____

2. _____

    _____

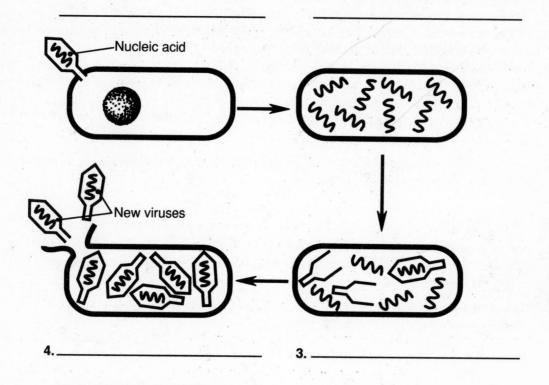

— Nucleic acid

New viruses

4. _____

    _____

3. _____

    _____

1. Why are viruses considered to be hostile to living things?

2. Describe a situation, if one exists, in which a cell would be a willing host to a virus. Explain your reasoning.

Reproduction of Viruses

Teaching Strategies

- Introduce the transparency by having the students recall that viruses display different shapes and affect living things in negative ways. Ask the students to examine the transparency and identify the material of which all viruses are composed. (All viruses are composed of nucleic acid.) Ask the students to describe the function of nucleic acid. (Nucleic acid is hereditary material, or a blueprint of genetic instructions.) Explain to the students that viruses are nonliving, noncellular genetic material surrounded by a protein coat. Have the students use the transparency to describe the sequence of events that enables a virus to reproduce.

- Ask the students what the cell invaded by a virus is called. (a host cell) Then distribute the worksheets for completion.

Questions

1. Why are viruses considered to be hostile to living things? (Viruses invade, then destroy, the cells of living things. An invaded cell serves as a host to one or more viruses. An invaded host cell abandons its normal function and uses the genetic material of the virus to create new viruses. The new viruses ultimately destroy the host cell by bursting from it and moving about to invade other cells.)

2. Describe a situation, if one exists, in which a cell might be a willing host to a virus. Explain your reasoning. (The students might suggest that since a virus alters the function of a host cell and ultimately destroys that cell, there is no logical situation in which a cell would be a willing host to a virus.)

Chapter 8

HRW material copyrighted under notice appearing earlier in this work.

9

INVESTIGATION

8.1   *Making Models of Viruses*

Purpose

■ To use information from a chart to make accurate scale models of several viruses

Materials

Metric ruler   Toothpicks
Plastic foam   Pipe cleaners
Clay   Wooden dowels
Insulated electrical wire   Tape
Cork   Construction paper

Procedure

1. Look at Table A. The illustrations in the first column show the basic shape of each virus. However, the drawings are not drawn to scale. That is, the real difference in the sizes of the viruses is not shown. Read the third column of Table A to find the actual size of each virus. Sizes are given in nanometers (nm). One nanometer is equal to 0.000 000 001 m. NOTE: Length values are not given for viruses that are more or less spherical in shape.

TABLE A: SIZES OF VIRUSES							
Shape of Virus	**Type of Virus**	**Actual size**		**Relative size**		**Scale size**	
		Diameter in nm	**Length in nm**	**Diameter**	**Length**	**Diameter in mm**	**Length in mm**
	Potato X	10	500	1.0	_____	5	250
	Polio	28		_____		14	300

(continues)

INVESTIGATION 8.1   *Making Models of Viruses* (continued)

		TABLE A: SIZES OF VIRUSES (continued)					
Shape of Virus	Type of Virus	Actual size Diameter in nm	Length in nm	Relative size Diameter	Length	Scale size Diameter in mm	Length in mm
	Influenza	100		10.0		50	___
	Mumps	200		20.0		100	___
	Tobacco Mosaic	18	300	___	30	9	___

2. The easiest way to find the relative sizes of the viruses is to assign them numbers that are easier to work with. The fourth column of Table A shows that the diameter of the Potato X virus has been assigned a relative size of 1. Notice that relative measurements have no units. This is the smallest virus in the table. To give the diameter of this virus a size of 1, you must divide its actual diameter by 10 nm:

$$10 \text{ nm} \div 10 \text{ nm} = 1.$$

3. Find the relative length of the Potato X virus by dividing its actual length by 10 nm. Record your result in the appropriate place in the fourth column.

4. To keep the relative sizes of the viruses the same, you must now perform the same procedure for each of the other viruses. Divide each diameter and length by 10 nm. Record your results in the appropriate places in the fourth column.

5. Before you can make your models of the viruses, you must decide on a scale size. You could use the relative sizes that you have already found. However, if you used the relative sizes as they are, some of your models might become too large to handle easily. You need to use a scale on which the smallest virus is easy to make, but on which the largest virus is not too large. For convenience, assume that the relative size of 1 equals 5 mm. To do this, multiply the relative diameter (1) by 5 mm. Then, do the same for the relative length: 50 × 5 mm = 250 mm. This gives the Potato X virus a scale size of 5 mm × 250 mm.

(continues)

INVESTIGATION 8.1 **Making Models of Viruses** (continued)

6. Follow the same procedure to find the scale sizes of the other viruses. Record your results in the fifth column of Table A.

7. Choose the best materials to use to make each virus. Make one scale model for each virus. Be careful to follow the scale sizes you determined in item 6.

Analyses and Conclusions

1. How can models such as those you have created help people understand the relative sizes of viruses?

_____

_____

_____

_____

2. What conclusion can you make about the relative sizes of the mumps virus and the influenza virus?

_____

3. Can you make a valid comparison of the size of the influenza virus and the Potato X virus? Explain your answer.

_____

_____

_____

Application

1. Review the information in your textbook regarding the way viruses enter cells and reproduce. In the space provided, write a paragraph explaining how the various shapes of viruses may help the viruses gain entry into cells and do their work.

_____

_____

_____

_____

(continues)

INVESTIGATION 8.1   *Making Models of Viruses* (continued)

_____

_____

_____

2. Write a paragraph discussing the variety of sizes in viruses. In your paragraph, suggest possible answers to the following questions:

   a. What relationship might the size of a virus have to the type of cell it invades?

   b. Is there such a thing as a typical virus? Explain your answer.

_____

_____

_____

_____

_____

_____

_____

_____

_____

_____

_____

_____

_____

_____

_____

_____

INVESTIGATION

8.2 *Observing Cyanobacteria*

Purpose

■ To study the characteristics of several types of cyanobacteria

Materials

Prepared slides of
   Gloeocapsa, Anabaena,
   Oscillatoria, and *Nostoc*
Compound light microscope

Procedure

1. Cyanobacteria are a group of organisms similar to many other bacteria. However, cyanobacteria contain chlorophyll, just as is found in green plants. Place a prepared slide of *Gloeocapsa* under a microscope. Observe the slide under low and high power. This type of cyanobacteria is unicellular. Note the circles that seem to surround each cell. These are actually layers of jellylike material. In the space provided, sketch the high-power view of the *Gloeocapsa*. Then, describe any distinguishing characteristics on the lines beside your sketch.

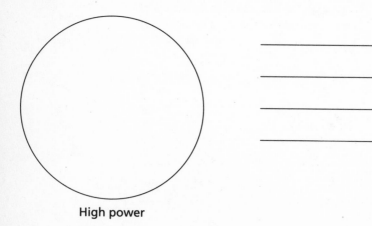

High power

_____

_____

_____

_____

(continues)

INVESTIGATION 8.2   *Observing Cyanobacteria* *(continued)*

Chapter 8

2. Place the slide of *Anabaena* under the microscope and observe it under low and high power. *Anabaena* is a multicellular cyanobacteria that forms long filaments, which are one cell thick. In the space provided, sketch a high-power view of *Anabaena*. Then, describe any distinguishing characteristics on the lines beside your sketch.

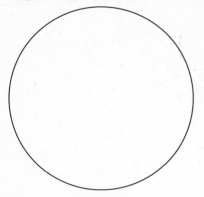

High power

_____

_____

_____

3. Place the slide of *Oscillatoria* under the microscope and observe it under low and high power. *Oscillatoria* is another cyanobacteria that forms long filaments. However, it is different in many respects from the *Anabaena* you studied in Step 2. In the space provided, sketch the high-power view of the *Oscillatoria*. Then, describe its characteristics on the lines beside your sketch. Be sure to include characteristics that distinguish *Oscillatoria* from *Anabaena*.

High power

_____

_____

_____

4. Place the slide of *Nostoc* under the microscope and observe it under low and high power. *Nostoc*, too, forms long, thin filaments. The arrangement of the filaments, however, differs from both *Anabaena* and *Oscillatoria*. In the space provided, sketch the high-power view of *Nostoc*. Then, describe the characteristics of Nostoc on the lines beside your sketch. Be sure to include characteristics that distinguish *Nostoc* from *Anabaena* and *Oscillatoria*.

(continues)

INVESTIGATION 8.2   *Observing Cyanobacteria* (continued)

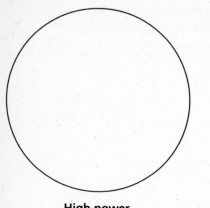

_____

_____

_____

_____

High power

Analyses and Conclusions

1. How is *Gloeocapsa* different from the other cyanobacteria?

   _____

2. What differences did you observe among the slides of *Nostoc*, *Oscillatoria*, and *Anabaena*?

   _____

   _____

Application

The growth rate of most kinds of cyanobacteria is directly related to the chemical content of the water in which it grows. Use reference sources to find out what **algal bloom** is. Write a short essay explaining what algal bloom is, its effect on the environment, and how it can be prevented.

_____

_____

_____

_____

_____

_____

_____

CHAPTER 8

Viruses and Monerans

Reading Skills

Making a Chapter Outline

Directions: Read Chapter 8. Then, on a separate sheet of paper, write a complex outline of the information in the chapter. A complex outline is one in which you include minor details as well as major ones. The format for a complex outline is shown below. Be sure to follow this format when you complete your outline.

Title of Outline

I. Main Idea 1
   A. Supporting detail for main idea 1
   B. Supporting detail for main idea 1
      1. Supporting detail for detail B
         a. Supporting detail for detail 1
         b. Supporting detail for detail 1
      2. Supporting detail for detail B
      3. Supporting detail for detail B
   C. Supporting detail for main idea 1
II. Main Idea 2
   A. Supporting detail for main idea 2
      1. Supporting detail for detail A
      2. Supporting detail for detail A
      3. Supporting detail for detail A
   B. Supporting detail for main idea 2
III. Main Idea 3
   A. Supporting detail for main idea 3—one supporting detail for detail A
   B. Supporting detail for main idea 3
      1. Supporting detail for detail B
      2. Supporting detail for detail B

    Notice that when a main idea or supporting detail is modified by only one detail, the detail is written on the same line as the idea it modifies (see III. A).

HRW material copyrighted under notice appearing earlier in this work.

17

CHAPTER
8

Viruses and Monerans

Science and Social Studies

Modern Medicine and Society

Infectious diseases have strongly influenced the course of civilization throughout human history. The Black Plague, for example, brought about the collapse of the feudal system of medieval Europe by killing off the majority of the working class. Major battles have been decided because armies were conquered by diseases, not by enemies. Infectious diseases have long dictated the areas in which people could settle. Even social contacts within a community can be affected by outbreaks of infectious diseases. The majority of the infectious diseases caused by bacteria can now be controlled through the use of antibiotics. Altered toxins, weakened live viruses, and dead organisms are all used today to immunize people against many life-threatening viral diseases.

A. Directions: Do research to find out how the control of diseases by using antibiotics and immunizations affects society. For example, when life-threatening diseases are controlled, people tend to live longer. How does this affect society in general? Then, on a separate sheet of paper, write an essay presenting the information you find.

B. Directions: In the space provided, answer the following questions.

1. List two infectious diseases that still cannot be controlled by antibiotics or immunization.

   _____

2. Why are the leading causes of death no longer the infectious diseases that killed so many people at the beginning of this century?

   _____

   _____

3. Scientists may someday be able to conquer all types of disease. What effect do you think this would have on society? Explain your answer.

   _____

   _____

   _____

18

CHAPTER

8

Viruses and Monerans

Extending Science Concepts

Calculating Bacterial Growth

Bacteria reproduce very rapidly. One bacterium can reproduce in about 20 minutes if conditions are favorable. After a bacterium reaches full size, it divides into two. The two bacteria then grow to full size and each divides into two again, forming four bacteria. If conditions are favorable, one bacterium can form millions of bacteria in a few short days.

To see how many bacteria can be formed in just six hours, design a table similar to Table A. The table should show for every 20-minute interval, the number of bacteria formed. Start with one bacterium at time 0:00.

TABLE A: BACTERIAL GROWTH			
Time	**Number of bacteria**	**Time**	**Number of bacteria**
0:20	_____	3:20	_____
0:40	_____	3:40	_____
1:00	_____	4:00	_____
1:20	_____	4:20	_____
1:40	_____	4:40	_____
2:00	_____	5:00	_____
2:20	_____	5:20	_____
2:40	_____	5:40	_____
3:00	_____	6:00	_____

(continues)

CHAPTER 8   *Viruses and Monerans* (continued)

Make a graph of the six hours of bacterial growth shown in Table A. Then answer the following questions.

1. How many cell divisions take place in one hour? In five hours?

   _____

2. Why do you think bacteria seldom ever multiply as rapidly as the table shows?

   _____

   _____

3. Explain the following statement: Bacteria multiply by dividing.

   _____

CHAPTER 8 Viruses and Monerans

Thinking Critically

Directions: Read about an experiment that Dr. Wilson performed. Then answer the questions about her experiment.

Dr. Erica Wilson has found a strain, or type, of bacteria called *Staphylococcus aureus* that she believes is causing a widespread infection among her patients. Dr. Wilson knows that there are many different strains of *S. aureus*. An antibiotic that kills one strain of the bacteria may not kill another strain. Therefore, Dr. Wilson set up an experiment to see which antibiotic she should use to fight the infection. This is the procedure she followed.

1. She obtained a pure sample of the strain of *S.aureus* she wanted to test.

2. She obtained a sterile Petri dish that contained **agar,** a jelled substance that provides food for the bacteria. The agar contained the same food sources that are available to bacteria inside the human body.

3. She spread a tiny amount of the bacteria onto the agar, being careful not to contaminate her sample or the agar.

4. She placed five antibiotic test discs onto the agar on top of the bacteria. The test discs she used were about 3 mm in diameter. Each test disc contained a different antibiotic in the amount usually given to people to fight infections.

5. On the bottom of the dish, she numbered each disc so she could tell later which one was which. (See Fig. A.) She made the following notes in her notebook:

Figure A
1 = Penicillin
2 = Streptomycin
3 = Tetracycline
4 = Ampicillin
5 = Sulfanilamide

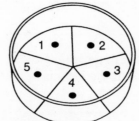

6. She covered the dish to avoid contamination. Then she placed it in an incubator (an ovenlike instrument that keeps a constant temperature). She set the incubator at 37°C (human body temperature).

(continues)

CHAPTER 8   *Thinking Critically* (continued)

7. After 48 hours, she looked at the dish. This is what she saw:

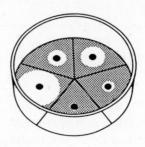

1. What might account for the fact that the bacteria did not grow around some of the test discs?

_____

_____

2. Why did the bacteria grow closer to some test discs than others?

_____

_____

3. Why was it necessary to make sure that the agar and the sample were not contaminated?

_____

_____

4. Ampicillin is derived from penicillin. Why do you think these two antibiotics showed different amounts of "clear zone"?

_____

_____

5. Judging from the results of the experiment, which antibiotic should Dr. Wilson give to patients to fight the infection?

_____

_____

_____

CHAPTER TEST

8 | *Viruses and Monerans*

Understanding Vocabulary

Explain how the terms in each pair are related.

1. host, nucleic acid of a virus

_____

2. AIDS, immune system

_____

3. antibiotic, vaccine

_____

4. virus, bacteria

_____

5. epidemic, time

_____

6. bacteria, toxins

_____

7. pigments, chlorophyll

_____

Understanding Concepts

MULTIPLE CHOICE

In the space to the left, write the letter of the choice that best completes the statement or answers the question.

_____ **8.** Viruses are able to reproduce
 a. anywhere in the environment of Earth.
 b. inside the living cells of plants.
 c. inside the living cells of animals.
 d. inside the living cells of plants and animals.

(continues)

HRW material copyrighted under notice appearing earlier in this work.

23

Chapter 8

CHAPTER TEST   *Viruses and Monerans* (continued)

_____ 9. Which of these activities could result in the transmission of the
AIDS virus?
a. shaking hands with an AIDS-infected person
b. breathing the same air as an AIDS-infected person
c. having sexual contact with an AIDS-infected person
d. having an AIDS-infected person as your friend

_____ 10. Which of these diseases is caused by bacteria?
a. strep throat   b. rabies
c. measles   d. influenza

_____ 11. Choose the statement that is most correct.
a. Bacteria and viruses exist in identical conditions.
b. Bacteria and viruses exist in similar conditions.
c. Viruses can exist in conditions that bacteria cannot exist in.
d. Bacteria can exist in conditions that viruses cannot exist in.

_____ 12. Cyanobacteria are an example of
a. viruses.   b. monerans.
c. pigments.   d. antibiotics.

Interpreting Graphics

13. These drawings show three typical bacteria. Describe the shapes of
typical bacteria.

_____

Reviewing Themes

14. *Systems and Structures*
Did scientists and researchers know that microscopic bacteria and
viruses existed before the invention of microscopes? Explain.

_____

_____

(continues)

CHAPTER TEST *Viruses and Monerans* (continued)

15. Environmental Interactions
What would happen to plants if cyanobacteria became extinct?

_____

_____

Thinking Critically

16. The shapes of human beings are similar in the sense that most have two legs, a head, and a trunk. Are the shapes of viruses also similar?

_____

17. Must one or more people die before a disease can be called an epidemic? Explain.

_____

_____

18. Bacteria reproduce very quickly. Why aren't more bacteria found on Earth?

_____

19. How can a person lessen the likelihood of acquiring the AIDS virus?

_____

20. Knowing what you know about viruses, is it possible for a person to contract a cold from exposure to cold weather?

_____

Performance Assessment

Observing Bacteria

Bacteria can be found almost everywhere in the world around you. However, many people have never seen what different bacteria look like. This investigation will allow you to observe various bacteria.

(continues)

Chapter 8

CHAPTER TEST *Viruses and Monerans* (continued)

THE MATERIALS

100-mL beaker • tap water • medicine dropper • microscope slide • cover-slip • plain yogurt or sour cream • methylene blue • compound light microscope

THE INVESTIGATION

Part A. **Readying the materials**

Fill a beaker with approximately 25 mL of water. Then add a small amount of plain yogurt or sour cream to the water and stir thoroughly. Using a medicine dropper, place a drop of the beaker mixture on a slide and stain it using a drop of methylene blue. Cover the slide with a cover-slip and mount it on a microscope. Remember to follow the proper procedure when using a compound light microscope.

_____

Part B. **Making your observations**

Using the different objectives of your microscope, locate bacteria.

_____

Part C. **Gathering your data**

Note the general shape and type of movement that the bacteria display. Then create a sketch of the different bacteria you observe, and include with each sketch a brief note describing the movement of the bacteria.

_____

_____

Part D. **Going beyond your investigation**

Recall that bacteria can be found almost everywhere in the world around you. Think of other substances or locations that you suspect may contain bacteria; then design and perform an experiment that will test for the presence of those bacteria.

_____

CHAPTER

9 Protists and Fungi

SECTION 1   Protozoans

A. Directions: In the space provided, write the letter of the term in the box that best fits the description.

_____ **1.** In amoebas, these structures collect extra water from the cytoplasm and release it through the cell membrane.

_____ **2.** These long, whiplike structures are used for movement in some protozoans.

_____ **3.** These move by using long, whiplike structures.

_____ **4.** These hairlike structures cover the entire cell and can create currents of water to bring food to the organism.

_____ **5.** All of these are parasites. They cannot move from place to place on their own.

_____ **6.** These are the most complex protozoans. They are covered with hundreds of hairlike structures.

_____ **7.** These projections made of cytoplasm are used for movement and to capture food.

_____ **8.** These protozoans move from place to place using pseudopods.

_____ **9.** This microscopic organism is single-celled and is most likely to be found in water. One kind causes malaria.

a. protozoan
b. pseudopods
c. cilia
d. flagella
e. contractile vacuole
f. sarcodines
g. ciliates
h. flagellates
i. sporozoans

B. Directions: Match each organism in Column A with the group of protozoans in Column B to which it belongs.

COLUMN A

_____ **1.** amoeba

_____ **2.** paramecium

_____ **3.** parasite that causes African sleeping sickness

_____ **4.** parasite that causes malaria

COLUMN B

a. ciliates

b. flagellates

c. sarcodines

d. sporozoans

Chapter 9

CHAPTER

9 Protists and Fungi

SECTION 2 Algae

Directions: Write the correct term from the list below to complete each sentence.

dinoflagellates	red tide	shells	*Euglena*
diatomaceous earth	diatoms	algae	fire algae
photosynthesis	pigments		

1. Often found in water and sometimes on tree trunks,

   _____ are members of the protist kingdom that have chloroplasts in their cells.

2. Algae use a process called _____ to make their own food from sunlight and nutrients.

3. Algae are classified according to the red, green, or brown

   _____ they contain.

4. One type of freshwater algae is the _____ , an organism that has characteristics of both plants and animals.

5. The _____ are a type of algae found in the ocean. Each organism moves using two flagella.

6. When dinoflagellates reproduce very rapidly and release poisons into

   ocean water, they can cause a _____ that kills fish.

7. Some dinoflagellates are nicknamed _____ because they glow like tiny fireflies when they are disturbed.

8. The _____ are golden brown algae that are the most common of all the single-celled organisms in the oceans.

9. The cell walls, or _____ , of diatoms are in two parts and made of a chemical similar to glass.

10. The empty shells of diatoms collect on the ocean floor to form

    _____ , a material used in many products.

CHAPTER 9

Protists and Fungi

SECTION 3 Fungi

A. Directions: Write one term from the list below for each description. Write each letter of the term on a separate line or box.

slime molds	spores	lichen
hyphae	fungi	mycelium

1. Organisms with bodies made up of hyphae that decompose organic material

    ____ ____ ☐ ____ ☐

2. Part fungus and part alga

    ____ ☐ ____ ____ ____ ☐

3. Move like amoebas, look and reproduce like fungi

    ____ ____ ☐ ____ ____

    ____ ____ ☐ ____

4. A group of interlocking hyphae

    ____ ____ ☐ ____ ☐ ____ ____ ____

5. Threadlike structures that make up the bodies of fungi

    ____ ____ ☐ ____ ____ ____

6. Formed by most fungi in order to reproduce

    ____ ____ ____ ____ ☐ ____

B. Directions: Unscramble the letters in the boxes to find the term for one of the most helpful members of the fungus kingdom.

_____

CHAPTER 9 *Protists and Fungi*

Chapter Review

Directions: Circle the letter of the best term for each description.

1. The protozoans are classified into four groups. Which of these is not one of those four groups?

   a. ciliates   **b.** parasites   **c.** sporozoans

2. Amoebas use these for movement and to capture food.

   a. pseudopods   **b.** cilia   **c.** flagella

3. Amoebas use contractile vacuoles for this function.

   a. store food   **b.** collect water   **c.** store waste

4. African sleeping sickness is caused by these organisms.

   a. tsetse flies   **b.** flagellates   **c.** flagella

5. Algae are important in ocean food chains because they can do this.

   a. make food   **b.** reproduce   **c.** decompose waste

6. These chemicals are used to classify algae.

   a. chlorophyll   **b.** chloroplasts   **c.** pigments

7. Which of these algae have two-part shells?

   a. dinoflagellates   **b.** *Euglena*   **c.** diatoms

8. Which of these algae live in fresh water?

   a. *Euglena*   **b.** diatoms   **c.** dinoflagellates

9. These organisms decompose organic material.

   a. algae   **b.** parasites   **c.** fungi

10. These organisms obtain food from dead organisms or waste products.

    a. parasites   **b.** saprophytes   **c.** algae

11. Which of these reproduce by forming spores?

    a. most fungi   **b.** most algae   **c.** protozoans

12. Lichens are not classified as fungi because they can do this.

    a. retain water   **b.** make food   **c.** move to a new place

INVESTIGATION

9   *Observing Protozoans*

(Textbook page 235) As you perform this investigation from your textbook, use this sheet to record your results and to answer the questions.

PROCEDURE

4. Using low power on your microscope, observe the drop of water. Draw and label what you see.

5. Switch to high power. Again draw and label what you see.

TABLE 1: OBSERVATION OF PROTOZOANS				
Protist Name	**Sketch**	**Description (Size/Shape/Color)**	**Type of Movement**	**Type of Feeding**

(continues)

HRW material copyrighted under notice appearing earlier in this work.

31

Chapter 9

INVESTIGATION 9   *Observing Protozoans* (continued)

ANALYSES AND CONCLUSIONS

1. What types of protozoans do you see? Flagellates? Ciliates? Amoebas?

   _____

   _____

   _____

2. Compare your drawings with those of your classmates. How are the organisms in your samples similar to and different from the organisms in their drawings?

   _____

   _____

   _____

   _____

   _____

3. How do the organisms from the different water samples compare?

   _____

   _____

   _____

APPLICATION

Why is it important for a scientist or a technician working in a water purification plant to be able to identify protozoans? Explain your reasoning.

_____

_____

1. _____

2. _____

3. _____

4. _____

5. _____

6. _____

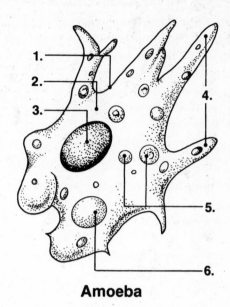

Amoeba

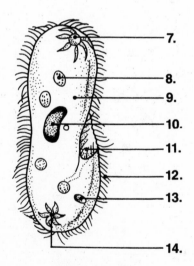

Paramecium

7. _____

8. _____

9. _____

10. _____

11. _____

12. _____

13. _____

14. _____

1. Why is it difficult to describe the shape of an amoeba?

2. Describe how an amoeba moves from place to place.

3. How do the cilia of a paramecium facilitate movement from place to place?

Protists

Teaching Strategies

- Introduce the transparency by asking the students to describe the shape of an amoeba and the shape of a paramecium. (The shape of an amoeba is difficult to describe because of its irregularity; a paramecium is elliptical in shape.) As you point to the structures of an amoeba and a paramecium, have the students identify those structures and their functions. Ask the students to explain why the shape of an amoeba changes frequently. (The movement of cytoplasm causes an amoeba to change shape.) Also ask the students to describe the function of the cilia which line the perimeter of a paramecium. (The cilia of a paramecium serve a locomotive function.)

- After reminding the students that amoebas and parameciums are animal-like in many ways and contain a variety of cell organelles, provide the students with worksheets for completion.

Questions

1. Why is it difficult to describe the shape of an amoeba? (The flow of cytoplasm constantly changes the shape of an amoeba.)

2. Describe how an amoeba moves from place to place. (Pseudopod projections made of cytoplasm, or "false feet," enable an amoeba to "creep" from place to place. During movement, the shape of an amoeba changes.)

3. How do the cilia of a paramecium facilitate movement from place to place? (Cilia are short hairlike structures that "sweep" or propel a paramecium through its environment.)

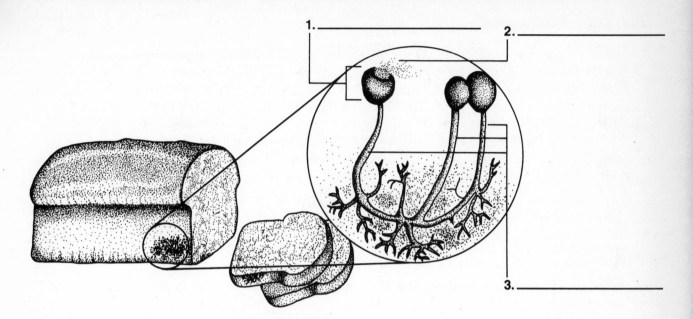

1. _____
2. _____
3. _____

Bread Mold

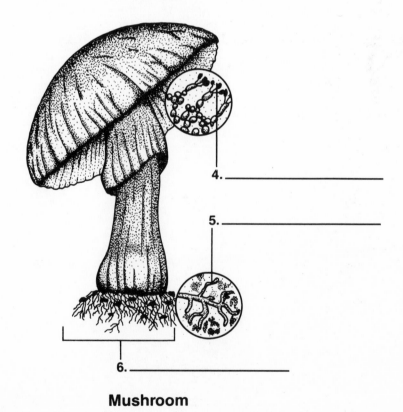

4. _____
5. _____
6. _____

Mushroom

1. What structure enables fungi to absorb food?

2. Is a living organism ever a willing host to fungi? Why or why not?

3. How do most fungi reproduce?

Fungi

Teaching Strategies

- Introduce the transparency by asking the students to describe the appearance and the location of any fungi they have seen in the outdoor environment. Remind the students that some forms of fungi, such as mushrooms, may be poisonous and *should not* be eaten. Then using the transparency, point to the structures of the mushroom and bread mold and ask the students to identify the structures and their functions. Also have the students recall that fungi cannot produce their own food—they rely on other organisms or materials in their environment for food.

- After asking the students to identify the reproductive structures of most fungi (spores), distribute the worksheets for completion.

Questions

1. What structure enables fungi to absorb food? (Hyphae; a network of hyphae is known as the mycelium of fungi.)

2. Is a living organism ever a willing host to a fungi? Why or why not? (Fungi usually injure or cause disease in a living host organism. As a result, organisms are likely to be unwilling hosts to fungi.)

3. How do most fungi reproduce? (Most fungi reproduce using spores.)

*I*NVESTIGATION

9.1   *Observing and Classifying Algae*

Purpose

■ To examine the characteristics of several different kinds of algae

Materials

Compound light microscope
Prepared slides of
   Chlamydomonas, Scenedesmus, Spirogyra, and *Volvox*

Procedure

1. Place a prepared slide of *Chlamydomonas* under the microscope. Examine the algae under low and high power. Sketch the organism.

2. Record the characteristics of *Chlamydomonas* in Table A. In the column labeled "Other Characteristics," list any characteristics you find interesting or that you think will serve to distinguish this organism from other algae.

3. Repeat steps 1 and 2 for *Scenedesmus, Spirogyra,* and *Volvox.* Be sure to label your sketches.

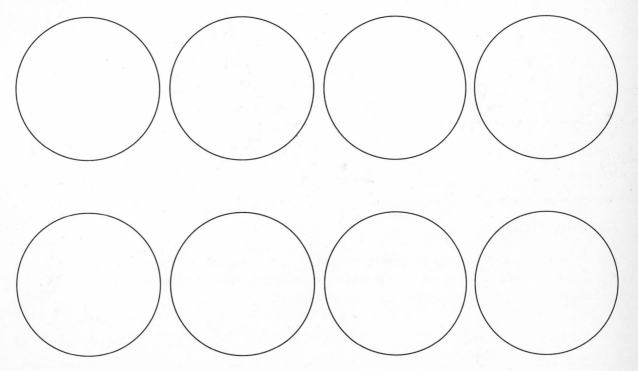

(continues)

INVESTIGATION 9.1   *Observing and Classifying Algae* (continued)

TABLE A: CHARACTERISTICS OF ALGAE				
Type of Algae	Shape of Algae	Unicellular or Multicellular	Flagella Present?	Other Characteristics
Chlamydomonas				
Scenedesmus				
Spirogyra				
Volvox				

Analyses and Conclusions

1. Which of the four types of algae do you think are able to move around freely? Explain your answer.

   _____

   _____

2. Of the multicellular algae that you examined, which do you think are the most likely to have specialized cells?

   _____

   _____

   _____

Application

All of the algae you studied in this investigation are freshwater algae. *Spirogyra* and *Chlamydomonas* live primarily in polluted water. How can knowing about these organisms help your community?

_____

_____

INVESTIGATION

9.2 *Observing the Growth of Yeast*

Purpose

- To determine the best growing conditions for yeast
- To observe the effects of pollution on the growth of yeast

Materials

Wax pencil
Test tubes (12)
Forceps
Active dry yeast
Granulated table sugar
Graduate, 15 mL

Distilled water
Test-tube racks (2)
Aluminum foil
Medicine dropper
Liquid dish detergent
Acetic acid solution, 5 percent

Procedure

Part A: Determining the best growing conditions for yeast

1. Using a wax pencil, label 8 test tubes with the numbers 1 through 8.

2. Using forceps, transfer 2 grains of dry active yeast into test tubes 3, 4, 7, and 8. Then transfer 4 granules of sugar to test tubes 5, 6, 7, and 8.

3. Add 5 mL of distilled water to all 8 test tubes. Roll each test tube gently between your hands to mix the contents.

4. Use Table A to check your test-tube setup.

TABLE A: TEST-TUBE SETUP					
Test Tube	**Distilled Water**	**Dry Yeast**	**Sucrose**	**Light**	**Dark**
1	×			×	
2	×				×
3	×	×		×	
4	×	×			×
5	×		×	×	
6	×		×		×
7	×	×	×	×	
8	×	×	×		×

(continues)

HRW material copyrighted under notice appearing earlier in this work.

39

Chapter 9

INVESTIGATION 9.2   *Observing the Growth of Yeast* (continued)

5. Place all the even-numbered test tubes in one test-tube rack, and place all the odd-numbered test tubes in the other rack.

6. Wrap the even-numbered test tubes loosely in foil to keep them in the dark, or store them in a dark cabinet. Leave the odd-numbered test tubes in a lighted area.

7. In Table B, note the appearance of the contents of test tubes.

TABLE B: APPEARANCE OF TEST TUBES		
Test Tube Set	**Beginning Appearance**	**Appearance after 24 Hours**
1		
2		
3		
4		
5		
6		
7		
8		

8. After 24 hours check the test tubes. In Table B, note the appearance of the contents of the test tubes.

Part B: Observing the effects of pollution on the growth of yeast

1. Label four test tubes A, B, C, and D. Determine which environment in Part A is best suited for growing yeast. Add those ingredients to all four test tubes. Test tube A will be a control.

2. Add two drops of liquid dishwashing detergent to test tube B.

3. Add two drops of 5 percent acetic acid to test tube C.

4. Add two drops of liquid dishwashing detergent and two drops of 5 percent acetic acid to test tube D. Fill in Table C to describe the contents of your test tubes.

5. Roll each tube gently between your hands to mix the contents.   *(continues)*

INVESTIGATION 9.2   ***Observing the Growth of Yeast*** *(continued)*

TABLE C: TEST-TUBE CONTENTS					
Test Tube	Distilled Water	Dry Yeast	Sucrose	Detergent	Acetic Acid
A					
B					
C					
D					

6. In Table D, note the appearance of each test tube.

TABLE D: APPEARANCE OF TEST TUBES		
Test Tube	Beginning Appearance	Appearance after 24 Hours
A		
B		
C		
D		

7. Allow the test tubes to stand for 24 hours in a light or dark environment, depending on your results in Part A.

8. After 24 hours, check the test tubes. In Table D, note the appearance of each set of test tubes.

Analyses and Conclusions

1. According to your results in Part A, what is the best environment for growing yeast?

_____

_____

(continues)

HRW material copyrighted under notice appearing earlier in this work.

41

Chapter 9

INVESTIGATION 9.2   *Observing the Growth of Yeast* (continued)

2. After performing the investigation in Part A, can you say that you have found the one best environment for growing yeast? Explain your answer.

_____

_____

_____

_____

3. In Part A, why do you think it was necessary to have one set of test tubes that contained nothing but water?

_____

_____

_____

4. Look at your results for Part B. What effect did the detergent have on the growth of yeast?

_____

_____

_____

5. What effect did the acetic acid have on the growth of yeast?

_____

_____

Application

Adding pollutants such as detergent and acid to a yeast culture made the yeast behave differently than it normally would. Explain what might happen to the water in lakes and rivers if large amounts of these pollutants were allowed to enter the water.

_____

_____

_____

CHAPTER

9 | *Protists and Fungi*

Reading Skills

Taking Notes

When you read a chapter, it sometimes helps to take notes about what you are reading. The procedure for taking notes is similar to that for making an outline. However, the procedure for taking notes is much less structured.

To take notes, first be sure you have the right equipment. You will need a pen or several sharpened pencils and some clean sheets of paper or a notebook. Then, as you read, write down ideas that you want to remember. These will usually include the main ideas and the major supporting details of a chapter.

Many people associate note taking with quickly scribbled, abbreviated notes that are almost unreadable. Although this is the way many people take notes, it is not the correct way. Effective notes must be easy to understand. There are many things you can do to make your notes easy to understand. Use the following guidelines to write notes that are good enough to use later to study for tests.

- At the top of the page, write the name and number of the chapter about which you are taking notes. This will help you identify the notes easily later.

- Always write clearly and neatly. Notes will do you little good if you cannot read them later.

- You may use abbreviations if you are sure that their meanings are absolutely clear to you. You must be able to go back to them later and know exactly what they mean. For example, the abbreviation *sub.* might stand for *substance*, *subject*, *subtopic*, or *substitute*. Abbreviations that can stand for more than one word may only confuse you later.

- Sometimes it helps to add a sentence or two to explain an idea more thoroughly. However, it is not necessary to write down every minor detail. Keep your notes as short and to the point as you can.

- After you have taken notes on the whole chapter, go back and review your notes immediately. If they seem clear, you are finished. However, if they seem disorganized, you may wish to rewrite your notes in what seems to you a more logical order. Add any details that are necessary to understand the notes.

(continues)

CHAPTER 9   *Reading Skills* (continued)

- After you have finalized your notes, keep them together in a note-book or other place where you can find them easily.

Directions: Reread Chapter 9 of your textbook. In the space provided, write notes about the chapter. Be sure to follow the guidelines discussed. If you think you need to rewrite or reorganize your notes, do so on another sheet of paper and attach it to this page.

_____

_____

_____

_____

_____

_____

_____

_____

_____

_____

_____

_____

_____

_____

_____

_____

_____

_____

CHAPTER
9

Protists and Fungi

Science and Language

Writing a Biographical Report

Dr. Alexander Fleming was born in Scotland and educated in London. He served as a professor of bacteriology from 1928 to 1948 at St. Mary's Hospital Medical School of London University. Dr. Fleming is recognized as an outstanding researcher in the fields of bacteriology, immunology, and chemotherapy. However, he is best known for a discovery he made accidentally in 1928, while working with a culture of staphylococcus [staf uh loh KAHK uhs], a type of bacteria. He noticed that a penicillium mold had stopped the growth of the bacteria. His discovery led to further discoveries and to the development of antibiotics.

Directions: Do research to find out more about the life and discoveries of Dr. Alexander Fleming. In the space provided, write a biographical report about Dr. Fleming. If you are unsure about the correct format for a biographical report, check with your teacher. If you need more space, attach another sheet of paper to this page.

_____

_____

_____

_____

_____

_____

_____

_____

_____

_____

_____

_____

Chapter 9

CHAPTER 9

Protists and Fungi

Extending Science Concepts

Antibiotics

You read in Chapter 9 that Alexander Fleming discovered the antibiotic properties of *Penicillium* mold. Fleming was experimenting with bacteria. Bacteria are commonly grown in flat, round dishes called *Petri dishes*. The dishes contain the nutrient agar in which bacteria can grow. Agar is a substance similar to gelatin. As the bacteria grow, the microscopic cells reproduce to form colonies, visible on the surface of the agar. Sometimes the bacterial colonies are so numerous that they cover the agar and make it appear cloudy. A clear area on the agar indicates no growth of bacteria. Bacterial colonies growing on nutrient agar are illustrated in Figure A.

Figure A

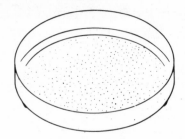

Petri dish with bacteria

As Fleming was observing his experiments, he noticed that some Petri dishes had become contaminated by mold. He thought his experiment was ruined and that he would have to throw out the dishes and start over again. Then he looked more closely and made a startling discovery. Bacteria were not growing near the mold. The mold seemed to stop, or inhibit, the growth of bacteria. The agar around the mold was clear—free of bacteria. Today we call this clear area a *zone of inhibition*.

Similar experiments can be used to test the effects of antibiotics on a variety of bacteria. Three molds—X, Y, and Z—are inoculated into 3 Petri dishes, each containing the same type of bacteria. Study Figure B and answer the questions that follow.

(continues)

CHAPTER 9 *Protists and Fungi* (continued)

Figure B

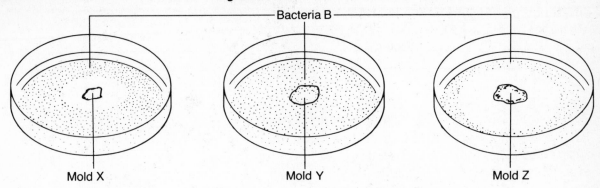

Three molds—X, Y, and Z—are inoculated into three petri dishes, each containing the same type of bacteria (B).

1. Which mold seems to have the best antibiotic properties against this kind of bacteria? Explain.

   _____

2. Which mold seems to have no antibiotic properties against this kind of bacteria? Explain.

   _____

3. An antibiotic was extracted from mold Z. How should its antibiotic effectiveness be tested?

   _____

   _____

(continues)

CHAPTER 9   Protists and Fungi (continued)

Antibiotic from mold Z is inoculated into 3 Petri dishes, each containing a different type of bacteria—Q, R, and S.

Figure C

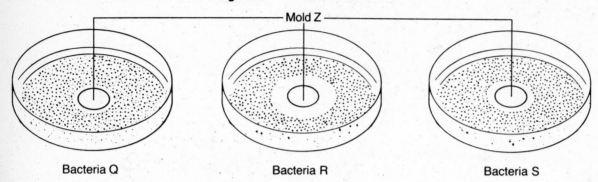

Bacteria Q   Bacteria R   Bacteria S

Antibiotic from mold Z is inoculated into three petri dishes, each containing a different type of bacteria—Q, R, and S.

4. Is antibiotic Z effective against all three types of bacteria? Explain.

_____

5. Is the antibiotic equally effective against all three types of bacteria? Explain.

_____

_____

6. Do you think antibiotic Z would be effective against all types of bacteria? Explain your answer.

_____

_____

_____

_____

_____

CHAPTER

9   *Protists and Fungi*

Thinking Critically

Directions: Over the last 20 years, the growth of algae in lakes, ponds, and streams has become a significant problem in many areas of the United States. Try to find an example of an algae-overgrowth problem in or near your community. If you cannot find evidence of such a problem in your community, you may choose a recent example of algae overgrowth that has been described in a recent newspaper or magazine. In the space provided, write an essay recommending a permanent solution to the problem. If you need more space, attach another sheet of paper to this page. As you write your essay, be sure to answer the following questions.

• What type of algae is causing the problem?

• What harmful effects are the algae having on the area?

• What is going into the water that causes the overgrowth?

• What efforts have been made by the community to overcome the problem?

• Has any progress been made in solving the problem?

• What other methods could be tried to solve the problem?

_____

_____

_____

_____

_____

_____

_____

_____

_____

_____

_____

Chapter 9

CHAPTER TEST

9

Protists and Fungi

Understanding Vocabulary

Write the letter of each organism in the blank next to the classification to which it belongs.

_____ 1. protozoan

_____ 2. alga

_____ 3. fungus

    a. *Euglena*

    b. *Penicillium*

    c. ciliate

    d. sporangium

    e. sarcodine

    f. flagellate

    g. kelp

    h. diatom

    i. sporozoan

    j. club

    k. dinoflagellate

    l. sac

Understanding Concepts

MULTIPLE CHOICE

In the space to the left, write the letter of the choice that best completes the statement or answers the question.

_____ 4. The means of locomotion used by an amoeba are
    a. pseudopods.   **b.** cilia.
    c. flagella.   **d.** vacuoles.

_____ 5. What is the function of the cell membrane of a protozoan?
    a. movement
    b. capturing food
    c. camouflage
    d. taking in oxygen and releasing carbon dioxide

_____ 6. Which of the following protozoans are always parasites?
    a. sarcodines   **b.** ciliates
    c. flagellates   **d.** sporozoans

(continues)

CHAPTER TEST   *Protists and Fungi* (continued)

_____ 7. Diatomaceous earth is the empty shells of a type of
        a. protozoan.   **b.** alga.
        c. fungus.   **d.** bacterium.

_____ 8. Which structure is similar in form to that of a flagella?
        a. tail   **b.** whip
        c. arm   **d.** leg

_____ 9. Which of these diseases is caused by sporozoans?
        a. amoebic dysentery   **b.** African sleeping sickness
        c. malaria   **d.** influenza

_____ 10. The mushroom is an example of a
        a. sac fungus.   **b.** imperfect fungus.
        c. club fungus.   **d.** sporangium fungus.

_____ 11. What characteristic is used to classify algae into different
        groups?
        a. the ability to perform photosynthesis
        b. the types of pigments contained by the algae
        c. the specific size of the algae
        d. the method of movement by the algae

_____ 12. Slime molds look like fungi, but they move like
        a. sarcodines.   **b.** algae.
        c. ciliates.   **d.** flagellates.

Interpreting Graphics

In the space, write the letter of the item that best completes each sentence.

_____ 13. The protozoan shown in this picture is a(n)
        a. amoeba.
        b. ciliate.
        c. flagellate.
        d. sporozoan.

_____ 14. The *Euglena* shown in this picture moves with the use of its
        a. nucleus.
        b. cell membrane.
        c. flagellum.
        d. chloroplasts.

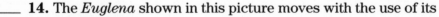

(continues)

Chapter 9

CHAPTER TEST   *Protists and Fungi* (continued)

Reviewing Themes

15. **Systems and Structures**
    What adaptations to land life do algae not need?

    _____

    _____

16. **Energy**
    Are microscopic organisms such as protozoans considered to be part
    of the food chain? Explain.

    _____

    _____

Thinking Critically

17. Do you think a sarcodine, a ciliate, or a flagellate would be more likely
    to lose a race? Why?

    _____

18. Why might it be difficult to describe the shape of an amoeba?

    _____

19. Even though algae are plantlike in many ways, algae lack several
    structures that are found on typical plants. Name three such structures.

    _____

20. Explain how algae can be both microscopic and macroscopic.

    _____

21. Why might you find fungi growing on a fallen tree in a forest?

    _____

22. How are lichens and slime molds similar to fungi?

    _____

    _____

(continues)

CHAPTER TEST   *Protists and Fungi* (continued)

Performance Assessment

Observing Fungi

Fungi are important organisms. Just as various animals look quite different from each other, fungi can look quite different from each other and display a wide range of shapes, sizes, and colors.

THE MATERIALS

dry yeast • sugar • methylene blue • 200-mL beaker • medicine dropper • warm tap water • microscope slide • coverslip • compound light microscope

THE INVESTIGATION

Part A. **Readying the materials**

Pour 100 mL of warm tap water into a beaker and add a large "pinch" of sugar. Add dry yeast to the mixture and allow it to stand covered in a warm place for approximately 30 minutes.

_____

Part B. **Performing the investigation**

Uncover the mixture and withdraw one or two drops for deposit on a slide. Stain the slide with methylene blue and use a coverslip to cover the slide. Then observe the slide using the different objectives of your microscope.

_____

Part C. **Gathering your data**

Sketch the fungi you see. On your sketch, label and identify as many fungi structures as you can.

Part D. **Communicating your results**

Compare your sketch with those of your classmates. As a class, suggest reasons why fungi occupy an important niche in Earth's environment.

_____

Chapter 9

UNIT TEST

3

Simple Living Things

⌐*Understanding Concepts*

MULTIPLE CHOICE

In the space to the left, write the letter of the choice that best completes the statement or answers the question.

_____ 1. Where are protozoans most likely to be found?
   a. in an ocean
   b. in tropical rain forests
   c. in the atmosphere
   d. underground

_____ 2. Which algae are the most common single-celled organisms in the oceans?
   a. *Euglenas*
   b. dinoflagellates
   c. diatoms
   d. fire algae

_____ 3. Choose the word that means the same as toxin.
   a. substance
   b. poison
   c. material
   d. disease

_____ 4. Which statement about monerans is correct?
   a. Monerans are never helpful to plants and animals.
   b. Monerans are sometimes helpful to plants.
   c. Monerans are sometimes helpful to animals.
   d. Monerans are sometimes helpful to plants and animals.

_____ 5. Protozoans are classified as
   a. plants.
   b. animals.
   c. protists.
   d. monerans.

_____ 6. The foods created by algae during photosynthesis provide energy for
   a. the algae.
   b. plants.
   c. animals.
   d. the algae and organisms that eat the algae.

(continues)

UNIT TEST   *Simple Living Things* (continued)

_____ 7. Where can viruses be found?
  a. Viruses can be found almost everywhere in the world around you.
  b. Viruses can only be found inside the living cells of plants.
  c. Viruses can only be found inside the living cells of animals.
  d. Viruses can only be found inside the living cells of plants and animals.

_____ 8. Which of these is *not* a locomotive structure of protozoans?
  a. pseudopods
  b. vacuoles
  c. cilia
  d. flagella

_____ 9. Which of these are responsible for epidemics?
  a. toxins
  b. bacteria
  c. viruses
  d. bacteria and viruses

_____ 10. Which part of a swimmer's body functions in a similar way to the cilia of a ciliate?
  a. lungs
  b. arms
  c. fingers
  d. toes

_____ 11. Why are illnesses caused by viruses difficult to treat and cure?
  a. Viruses are not living organisms.
  b. Viruses are difficult to locate.
  c. Viruses are too small to be seen.
  d. Viruses move from one place to another very rapidly.

_____ 12. In terms of movement, a dinoflagellate is most like which protozoan?
  a. a sarcodine
  b. a flagellate
  c. a ciliate
  d. a sporozoan

_____ 13. Which of these diseases is caused by viruses?
  a. tetanus
  b. diphtheria
  c. influenza
  d. strep throat

(continues)

Unit 3 Test

UNIT TEST   **Simple Living Things** (continued)

_____**14.** Which of these protozoans cannot move from one place to
        another on their own?
        a. sarcodines
        b. ciliates
        c. flagellates
        d. sporozoans

_____**15.** Which protozoans are most complex?
        a. sarcodines
        b. ciliates
        c. flagellates
        d. sporozoans

Interpreting Graphics

16. Identify each of these protozoans and their locomotive structures.

_____

_____

Reviewing Themes

17. *Systems and Structures*
    Explain how a lichen functions as a fungus and as an alga.

_____

_____

18. *Environmental Interactions*
    Explain why a fungus that is a saprophyte is important to a woodland
    environment?

_____

_____

(continues)

UNIT TEST *Simple Living Things* (continued)

Thinking Critically

19. Name any organism that would not be able to exist without proto-zoans, and describe how protozoans contribute to the life of that organism.

_____

_____

20. Some scientists do not agree that algae should be classified as protists. Is such disagreement healthy for the scientific community?

_____

21. Viruses can only reproduce inside the cells of living things, but bacteria can reproduce throughout the environment. Do you think a hiker in the woods would be more likely to acquire a bacterial infection than a viral infection? Explain.

_____

_____

22. Have viruses ever affected you? If so, how?

_____

23. Do you think scientists and researchers will ever discover a vaccine for all known viruses? Why or why not?

_____

24. Why is it said that people who have AIDS die indirectly from the virus?

_____

Unit 3 Test

Laboratory Safety for the Teacher

Laboratory safety is an important consideration for any science teacher. In order to create a safe laboratory, the teacher must drill the students on safety procedures. The students, of course, are responsible for themselves to some extent, but teachers have the ultimate responsibility for making sure proper procedures are enforced.

Many states provide some statutory framework related to school and laboratory safety that includes building design, safety equipment, and chemical use and disposal. Specifics of your local guidelines should be available through the school administration. In the following discussion, several important concerns are presented. However, the Pupil's Edition of the textbook has a substantial list of safety guidelines that you should review with your classes.

Laboratory areas should be provided with the following safety equipment:

Fire extinguisher	First-aid kit
Fire blanket	Laboratory aprons
Shower	Non-sterile latex gloves
Eyewash	Thermal mitts
Laboratory hood	Safety goggles

Fires of all types can be extinguished with the CO_2 fire extinguisher generally provided in most schools. Fires of paper or wood products may also be extinguished with water. For electrical fires and chemical fires, only the CO_2 fire extinguisher should be used. A fire blanket should be used to wrap around a student if his or her clothes or hair should catch on fire. Care should be taken to avoid the use of open flames whenever there is a viable alternative, such as a hot plate. The students should be required to restrain loose clothing or long hair when working in the laboratory.

Showers are for use when chemicals have been spilled on a student or on a student's clothing. The student should stand under the shower until the chemical is totally diluted. Occasionally, a student's clothing will be so contaminated, the clothing will have to be removed. It is a wise idea to have some replacement clothing handy should this circumstance occur; a robe or coveralls work especially well.

Eyewashes are to be used when chemicals are splashed onto the face or into the eyes. The students should leave the exposed area in the eye-wash for five to ten minutes.

Laboratory hoods are recommended whenever the students are working with highly volatile or noxious chemicals. If no laboratory hood is available, make sure the area is well ventilated.

Every laboratory area should have a first-aid kit. A typical kit contains an assortment of antiseptics, bandages, gauze pads, and scissors. Most kits also come with simple instructions for use. Be sure to read these instructions if you are not familiar with basic first-aid procedures. Be sure to take the first-aid kit on all field trips. In addition to first-aid procedures, it is helpful to be familiar with CPR (cardiopulmonary resuscitation).

Laboratory aprons and surgical gloves should be available and should be used by the students when working with toxic or potentially toxic chemicals.

Safety goggles or face protectors should be available and should be used by the students when working with volatile or irritating chemicals.

All reagents for student use should be clearly labeled and stored in secure areas. Reagents or chemicals not ordinarily used by the students should be kept in a locked area.

Spill kits for acids, bases, and mercury are available from supply companies. These kits allow for neutralization and easy cleanup of potentially dangerous chemicals. Baking soda (sodium bicarbonate) may also be used to neutralize acid spills.

HRW material copyrighted under notice appearing earlier in this work.

59

Do not dump dangerous wastes down the sink or into the trash. Dispose of them according to appropriate state or local guidelines.

Find out your school's policy and procedure for handling emergencies. Talk to the school nurse if your school has one. You should be familiar with the appropriate emergency procedures. If an accident does occur, remain calm. Encourage the students to be quiet and remain in their places. Handle the situation quickly, but calmly.

Solutions and Media

Preparation of Solutions

In scientific experiments, it is often necessary to make solutions of varying concentrations. Several different methods of determining concentration exist. Solutions may be prepared by percentage of solute, molarity, molality, and normality. For the purpose of simplicity, this laboratory manual employs only the percentage technique.

Percentage Solutions

Percentage by Volume (liquids)

To calculate percentage by volume, first decide the total amount of solution that is needed. For example, 1 L of solution may be necessary to complete an experiment for five classes of 30 students each. Then multiply the total volume needed by the percentage of the solution you desire. For example, if you need a 5% solution of hydrochloric acid, you can determine the amount of acid to use by doing the following calculation: 1 L × 5% equals 50 mL of hydrochloric acid. Subtract the amount of acid from the total amount of solution (1000 mL – 50 mL = 950 mL). The result is the amount of water needed to prepare the solution. Add the 50 mL of acid to the 950 mL of water, and a 5% HCl solution is the result.

Percentage by Mass (solids)

Calculating percentage by mass is similar to calculating percentage by volume. However, instead of using mL, you must measure grams. Suppose, for example, that you need about 1 L of a 5% sodium chloride solution. First you would calculate how many grams of NaCl you would need to produce a 5% solution. To do this, you must multiply 1000 g × 5% = 50 g of NaCl. (It is possible to change from mL to g with water because the density of water is approximately 1 g/cm^3, or 1 g/mL.) Measure 50 g of NaCl, pour it into a graduated cylinder, and then add 950 mL of distilled water.

Hazardous Animals and Plants of North America

Animals

It is often useful to have a list of potentially dangerous animals. However, it is not possible to fully describe the unique animals that are present in every area in such a general list. You can get a complete list about your area from your county extension agent.

Venomous Terrestrial Invertebrates

1. Wasps All wasps are predators and carry a potent venom to subdue prey. The venom itself usually causes no more than a stinging or localized pain, but some people are extremely allergic to the venom.

2. Velvet ants Velvet ants are actually wingless wasps. Like all wasps, they have venom and a stinger at the tip of their abdomens. Velvet ants are found on the ground and usually sting only if stepped on.

3. Widow spiders Several varieties of widow spiders occur in North America; the most common and famous of which is the *black widow*. Black widow spiders will not bite unless they are touched. Like all spiders, they have fangs and venom. Unlike most spiders' venom, widow venom can be lethal to humans. Venom from a single bite may be lethal to a small child. Multiple bites may be lethal to an adult. However, antivenin is available at medical facilities.

4. Brown recluse spiders Brown recluse, or fiddle-back, spiders do not breed over most of North America. However, they are often carried in cars, trucks, or other conveyances and may be encountered almost anywhere. Although it is not normally lethal, the venom can cause a painful sore that may take months to heal.

5. Scorpions North American scorpions are seldom a problem to humans. However, certain species are large and have potent venom. All scorpions should be handled carefully or avoided completely since they have a sharp stinger at the tip of their abdomen.

6. Fire ants Fire ants are found mostly in the southeast, but are spreading through the southwest also. Each ant is only a few millimeters long but has a potent venom that causes intense, burning pain and an allergic reaction in some people.

7. Bees Honeybees and their larger relatives, bumblebees, are not predators and therefore do not have very potent venom. However, some people may have an allergic reaction to the venom.

Annoying Terrestrial Invertebrates

In the following list are animals that seldom cause any serious problems but can produce annoying itching or stinging.

Wheel bugs	Common ants
Mosquitoes	Chiggers
Horseflies	Blister
Deer flies	beetles
Io moth caterpillars	Ticks*
Puss caterpillars	Fleas
Large spiders	Centipedes

*Note: Certain ticks carry a type of bacteria that causes Lyme disease in humans.

Venomous Terrestrial Vertebrates

1. Coral Snake Coral snakes produce strongly neurotoxic venom. They are found in the southern United States, Mexico, Central America, and parts of South America. Their bodies are brightly colored with encircling bands of red, yellow, and black. Bites usually occur on hands or feet because this snake has a small mouth, and short, nonerectile fangs. Nevertheless, coral snakes should be regarded as dangerous.

HRW material copyrighted under notice appearing earlier in this work.

61

2. Pit-viper snakes *Rattlesnakes*—Several species of rattlesnakes live in North America. Adults range in size from less than a foot to several feet in length. All rattlesnakes are pit vipers, as are the cottonmouths and copperheads and all are dangerous. Generally, all pit vipers have hollow, erectile fangs that act like hypodermic needles to inject venom. Although deaths from rattlesnake bites are rare, the bite of any pit viper is serious. Antivenin is available for all species and the bites can usually be neutralized by prompt medical treatment. **Under no circumstances should students catch or keep venomous snakes!**

Cottonmouths—Cottonmouths, or "water moccasins" as they are sometimes called, are aquatic, heavy-bodied pit vipers. They are usually aggressive and can inflict a painful bite. Although the venom is seldom lethal, the bite is dangerous.

Copperheads—Copperheads are upland relatives of the cottonmouth. They are in the same genus and their venom is similar.

3. Gila monster The Gila monster is the only venomous lizard in the United States. This lizard is large-bodied and brightly colored. It is not a threat to humans unless handled. Gila monsters are protected by law in Arizona, where they commonly live, and should not be disturbed.

4. Giant marine toad The giant marine toad is established in parts of Texas and south Florida. It resembles common toads, but it is much larger—20 to 25 cm long. These toads, like all toads, have poison glands on their backs. Unlike other toads, however, the toxin of the giant marine toad is quite potent and can even kill a small dog. If the toad is handled, the venom can get on one's hands and easily cause burning irritations if rubbed on the face or eyes. Giant marine toads should be avoided.

Venomous Marine Animals

1. Jellyfish All true jellyfish have stinging cells, which they use to capture prey. Some species, such as the Portuguese man-of-war, have venom of sufficient potency to cause harm to humans.

2. Fire coral Fire coral has potent stinging cells that can cause intense burning pain.

3. Stingrays Several species of stingrays occur in shallow coastal waters in North America. Many have a stinging spine located at the base of the tail. This spine is associated with a venom gland and can produce not only a painful injury but also an allergic reaction.

4. Toadfish or scorpion fish Certain fishes have venom sacs associated with dorsal fin spines. Toadfish and scorpion fish are examples. As with the stingray, careless handling can produce a painful cut and a potentially serious reaction.

5. Puffer fish Puffer fish have toxin in their tissues. Many species are lethal if eaten. In addition, some forms also have sharp spines that can cause painful wounds.

6. Reef fishes Several species of reef fishes are harmful and contain toxin that can cause illness if ingested. Such species include the butterfly fish, parrotfish, and surgeonfish. Often these fish accumulate toxins from the stinging cells of invertebrates in their diets.

Animals That Cause Mechanical Harm

Several species of animals can cause cuts to hands or feet if handled or stepped on. Such animals include both freshwater and marine catfish, oysters, and clams. Any animals that have sharply pointed body parts, such as spines or shells, should be treated carefully. Many vertebrates have teeth, especially those small mammals commonly kept in the classroom. **Caution should be exercised when handling any small rodent.**

Plants

It is difficult to generalize about the potential for plant poisoning, since plants contain such a wide variety of toxic, or potentially toxic, chemicals. In fact, most plants can produce toxic effects if ingested in large enough quantities. Detailed information about the most dangerous plants in your area is available from your county extension agent.

Tips on Avoiding Plant Poisoning

1. Become familiar with dangerous plants in your area.
2. Never eat wild plants or fungi.
3. Learn to recognize the most common poisonous plants, such as poison ivy.
4. Avoid smoke from burning plants.
5. There are no safe or reliable tests, or rules of thumb, to distinguish safe plants from poisonous ones.

Wild Plants That Cause Dermatitis

Brazilian pepper	Poison sumac	Poisonwood
Poison Ivy	Poison oak	Stinging nettle

Wild Plants That Cause Internal Poisoning

Beech	Elephant ear	Pokeweed
Black cherry	Holly	Virginia creeper
Buckeye	Jimsonweed	Water hemlock
Buttercup	Larkspur	Yellow jessamine
Chinaberry	Mistletoe	Yew
Coontie	Mushrooms	
Cycads	Nightshade	

Cultivated Plants That Cause Internal Poisoning

Angel's trumpet	English ivy	Rhododendron
Azalea	Foxglove	Rhubarb
Bleeding heart	Holly	Tobacco
Boxwood	Hyacinth	Trumpet flower
Caladium	Jessamine	Tung oil tree
Castor bean	Lantana	Wisteria
Cherry	Lignum vitae	Yew
Chinaberry	Oleander	Yellow jessamine
Crape jasmine	Pencil tree	
Cycads	Poinsettia	
Dieffenbachia	Privet	

Safety, Materials, and Notes

Maintaining Animals and Plants in the Classroom

Animals

What kinds of animals are suitable for classroom pets? A good rule to follow is keep animals that are normally kept as house pets. Visit a large pet store to get ideas. A variety of fishes, small mammals, small birds, and some reptiles are routinely kept. For example, rat snakes are commonly available in pet shops and are popular with students. Under no circumstances should poisonous or venomous species be collected or kept. Regardless of what types of animals you may decide to keep in your classroom, a few simple rules should be followed to ensure the animals' safety and health.

Terrestrial Animals

1. Animals should be kept in easy-to-clean cages that lock securely. Most teachers who keep animals find that their classrooms become popular places for students to come visit from other classes. Do not assume that students will not harass or handle the animals. Ensure the safety of the animals by keeping cages locked. When keeping small mammals or snakes, an all-glass aquarium with a securely fitting top makes a good cage. Regular cleaning of cages is necessary to keep down parasites and promote the good health of the animal.

2. Provide the animals with an appropriate diet. If you do not know what to feed a particular animal, seek advice from your local pet shop owner or veterinarian.

3. Determine the temperature needs of each animal. Avoid placing cages in direct sunlight. Animals may overheat if they cannot move out of the sunlight when they get too hot. Also, some areas may be too cool for certain animals. For example, boa constrictors are common snake pets, yet they require warm temperatures and are subject to respiratory infections if they are kept too cool.

Fish

1. Tap water is not suitable for fish since the chlorine in most tap water will kill fish. There are several ways to eliminate chlorine from tap water. You may let the water stand for a day or two. Chlorine dissipates as the water stands. If you do not have time to age the water, you may use a commercial chlorine remover available at most pet stores.

2. Another problem with fish is overfeeding. Fish eat only a limited amount of food at a time. If the fish are habitually overfed, the excess food fouls the water. Have a regular feeding time for your fish. Keep to this schedule and problems will be avoided.

3. Aquaria need to have regularly scheduled water exchanges to keep down the concentration of wastes. You can exchange the water by using a siphon. While old water is being siphoned out, debris can be sucked off the bottom of the tank at the same time. You should never exchange more than about 1/3 of the water in a tank at one time.

4. Aquaria need a good filtration and aeration system for good fish maintenance. The most reliable type of filter is the undergravel type. Sizes to fit any aquarium are available at most pet shops.

Plants

Plants are generally easier to maintain in the classroom than are animals.

1. Plants' major need is water. Often, plants are either over-watered or under-watered; so the watering should be done on a regular schedule. Most commonly kept houseplants will thrive in the classroom.

2. If you do not have windows, fluorescent light usually provides sufficient light for good plant growth. You may wish to keep all the plants in one area of the room and use a timer so the lights will still come on when you are gone on weekends.

Before you decide to keep plants and animals in the classroom, check the list of hazardous animals and plants found on the preceding pages.

Materials Lists

The materials lists have been compiled to help the teacher gather and order supplies for *Life Science* Laboratory Investigations. **Note: The *Annotated Teacher's Edition* of the textbook includes a listing of suppliers in the Additional Materials and Resources section.**

Two lists are provided. The Master Materials List identifies materials needed for all Laboratory Investigations. It organizes the materials in these categories: Apparatus and Equipment, Biological Material, Chemicals and Reagents, Glassware, and Local Supply. Each item in the list is keyed to the investigation for which it is needed. The second list identifies the materials needed for each investigation in this unit.

Master Materials List

Apparatus and Equipment	Investigations	Apparatus and Equipment	Investigations
Aquarium	15.2, 17.2	Incubator	14.2
Balance	1.1	Laboratory apron	1.2, 5.1, 5.2, 6.1, 12.1, 12.2, 13.1, 13.2, 14.1, 17.1
Compound light microscope	2.1, 2.2, 4.1, 5.2, 7.1, 8.2, 9.1, 10.1, 10.2, 17.2, 19.2	Lens paper	2.1
Cork	4.1, 8.1	Magnifying glass	12.2, 14.2, 18.2, 20.2
Culture tubes and caps, 20 mm x 200 mm	3.2	Metal probe	18.2
Dialysis tubing	5.1	Metric ruler	5.1, 7.2, 8.1, 16.2
Dip net	15.2, 17.2	Meter stick	16.2
Dissecting needles	13.1, 13.2	Ring stand and ring	2.2, 11.1
Dissecting pan	12.2, 13.1, 13.2, 14.1	Safety goggles	5.1, 5.2, 17.1
Dissecting pins	13.1, 14.1	Scalpel	14.1
Dissecting scissors	13.1, 13.2, 14.1	Scissors	2.1, 5.1, 5.2, 16.2, 20.1
Filter paper	2.2	Surgical gloves	12.1, 12.2, 13.1, 13.2, 14.1
Forceps	9.2, 10.1, 14.1	Test-tube holders	17.1
Hand lens	10.1, 11.2, 12.1, 13.1, 13.2, 14.1, 15.1	Test-tube racks	9.2
Hot plate	3.1, 5.1, 6.1, 15.2, 17.1, 20.1	Trays, aluminum	10.2, 12.1

Biological Material	Investigations	Biological Material	Investigations
Beef kidney	18.2	*Oscillatoria*	8.2
Cones from a pine tree Pollen cones	10.2	*Scenedesmus*	9.1
Seed cones	10.2	Skin cells	4.1
Chicken eggs, fertilized	14.2	*Spirogyra*	9.1
Chicken kidney	18.2	*Volvox*	9.1
Crickets or grasshoppers	12.1	whitefish blastula	5.2
Daphnia culture	19.2	Preserved specimens Annelids (segmented worms)	12.2
Earthworms	12.1	Arthropods	12.2
Elodea, live	3.2	Butterflies	15.1
Feathers	12.1	Chordates	12.2
Ferns, several kinds	10.1	Coelenterates	12.2
Freshwater snails	3.2	Earthworms	13.1
Frog eggs, fertilized	20.2	Echinoderms	12.2
Goldfish	15.2, 17.2	Flies	15.1
Hog kidney	18.2	Frogs	14.1
Mosses with spore cases, several kinds	10.1	Grasshoppers, female	13.2, 15.1
Prepared slides *Anabaena*	8.2	Mosquitoes	15.1
Blood cells	4.1	Mullusks	12.2
Bone cells	4.1	Nematodes (roundworms)	12.2
Chlamydomonas	9.1	Platyhelminthes (flatworms)	12.2
Gloeocapsa	8.2	Starfish	12.1
Muscle cells	4.1	Snails	12.1
Nerve cells	4.1	Sow bugs	12.1
Nostoc	8.2		

Chemicals and Reagents	Investigation	Chemicals and Reagents	Investigation
Acetic acid solution	9.2, 19.1	Phenolphthalein solution	18.1
Acetocarmine stain	5.2	Quinine sulfate solution	19.1
Alcohol solution	19.2	Salivary amylase	17.1
Benedict's solution	5.1, 17.1	Sodium bicarbonate solution	19.1
Bromothymol blue solution	3.2	Sodium chloride solution	19.1
Caffeine solution	19.2	Sodium hydroxide solution	18.1
Glucose solution	19.1	Starch solution	17.1
Glycerin	10.1	Sucrose solution	19.1
Methyl cellulose	7.1	Sugar solution	5.1, 17.1

Glassware	Investigations	Glassware	Investigations

Beakers
  100 mL
	19.2	Graduated cylinders
		  10 mL
			5.1

150 mL	2.2, 20.1	15 mL	9.2
250 mL	6.1, 17.1	100 mL	1.1, 1.2, 3.1, 6.1, 15.2
500 mL	1.1, 1.2, 3.1, 5.1, 15.2	Medicine droppers	2.1, 7.1, 9.2, 18.1, 19.2
Concave slides	19.2	Microscope slides	2.1, 2.2, 4.1, 5.2, 7.1, 10.1, 10.2, 17.2
Coverslips	2.1, 2.2, 4.1, 5.2, 7.1, 10.1, 10.2, 19.2	Petri dishes	14.2, 17.2, 18.2, 20.2
Flasks, 125 mL	18.1	Test tubes	5.1, 9.2, 17.1
Funnels	2.2, 5.1	Thermometers	3.1, 15.2, 20.1, 20.2

Local Supply	Investigations	Local Supply	Investigations
Acorns	7.2	Flowerpots, porous clay	10.1
Active dry yeast	9.2	Fruits: peach, apple, melon, pepper, maple, hickory, oak, green bean, pea pod, ear of corn	10.2
Adhesive tape	17.1	Glue, fast drying	4.2, 6.1
Aluminum foil	1.1, 9.2	Graters	3.1
Aluminum pie tin	6.1	Hacksaw	4.2
Aspirins	19.1	Hay	3.1
Beans dried	4.2	Hole punch	16.2
red	6.2	Hot pads	6.1, 20.1
seeds	11.1	Ice	15.2
white kidney	6.2, 7.2	Insulated electrical wire	8.1
Brads	16.2	Jars Clear glass	12.1
Buttons	4.2	Wide-mouth	10.1
Cans, tin	20.1	Knives, vegetable	10.2
Candles	5.2	Laundry detergent, powdered, with a high-phosphate content	1.1
Cardboard	16.2, 20.1	Leaves from vascular plant	6.1, 10.1
Cereal flakes	6.1	Lids, tin	6.1
Clay	6.1, 8.1	Lemon juice	19.1
Corn syrup, light	6.1	Lettuce, dried leaves	3.1
Cotton, absorbent	17.2, 20.1	Maple leaves	7.2
Cotton swabs, sterile	19.1	Marking pens	11.1, 17.1
Dishes	10.1	Masking tape	1.1, 3.2, 11.1
Dishwashing detergent, liquid	3.1, 9.2	Matches	5.2
Dried grass	3.1	Mixing bowls	1.2
Electric or hand mixer	1.2	Mixing containers	6.1

Notebooks	11.2, 15.1	Seedlings, bean, in small pots	11.1
Nuts	4.2	Shells	6.1
Onion sets	5.2	Soda bottle, 2 L, clear plastic	4.2
Paper Construction	8.1	Stopwatch	12.1, 19.2
Newspaper	1.2, 2.1	Straws	18.1
Towels	1.2, 2.1, 5.2, 10.1, 11.1, 12.2, 13.1	String or twine	5.1, 6.1, 16.2
White	10.1	Sugar, granulated table	6.1, 9.2
Paper clips, large	6.1, 16.2	Tape, clear plastic	4.2, 8.1, 17.2
Pencils	6.1	Thread	4.2
Pens or pencils in 4 colors	7.2, 20.1	Tongue depressors	6.1
Petroleum jelly	6.1	Toothpicks	7.1, 8.1
Pine needles	7.2	Vegetable slices	4.2
Pipe cleaners	8.1	Vinegar	9.2
Plaster of Paris	6.1	Wallpaper paste (powder)	1.2
Plastic bags, small zippered variety	11.1	Watch or clock with second hand	15.2, 17.2, 18.1
Plastic bubble wrap	20.1	Water Aquarium, well oxygenated	17.2
Plastic containers, 2 L	3.1, 6.2	Dechlorinated tap	3.2
Plastic foam	8.1	Distilled	2.1, 3.1, 4.1, 6.1, 9.2, 10.1
Plastic wrap	3.1, 11.1	Pond	1.1, 2.2, 3.1
Polystyrene cup	6.1	Wax paper	1.2
Polystyrene pieces	20.1	Wax pencil	3.1, 6.2, 9.2
Pond culture	2.1, 7.1	Weight, heavy, large book or bricks	1.2
Poster board	4.2	Wheat grains	3.1
Push pins	11.1	Wooden clothespins	5.2
Rice grains	3.1	Wooden dowels	8.1
Rubber bands	11.1	Wool	20.1
Screen, wire window, 10 cm x 10 cm	1.2		

Materials List for Investigations

INVESTIGATION 8.1

Metric ruler
Plastic foam
Clay
Insulated electrical wire
Cork
Toothpicks
Pipe cleaners
Wooden dowels
Tape
Construction paper

INVESTIGATION 8.2

Prepared slides of
  Gloeocapsa, Anabaena,
  Oscillatoria, and *Nostoc*
Compound light microscope

INVESTIGATION 9.1

Compound light microscope
Prepare slides of
  Chlamydomonas, Scenedesmus,
  Spirogyra, and *Volvox*

INVESTIGATION 9.2

Wax pencil
Test tubes (12)
Forceps
Active dry yeast
Granulated table sugar
Graduated cylinder, 15 mL
Distilled water
Test-tube racks (2)
Aluminum foil
Medicine dropper
Liquid dish detergent
Acetic acid solution, 5 percent
  (vinegar works well)

*Safety, Materials,
and Notes*

HRW material copyrighted under notice appearing earlier in this work.

73

INVESTIGATION 8.1

Pre-Lab

Have the students read the entire investigation before they begin working. Ask the students to examine the chart and to identify the viruses whose models, they think, will have the smallest diameter and the shortest length.

Hints

Before students begin this investigation, make sure that all students understand the mathematical value of μm. You may wish to do the calculations for the Potato X virus as an example.

Post-Lab

Ask the students to determine whether they accurately identified the viruses with the smallest diameter and shortest length. Ask the students to discuss why the variety of sizes and shapes might make identifying viruses difficult.

INVESTIGATION 8.2

Background

Like other monerans, cyanobacteria lack distinct nuclei. Some are single-celled organisms; others have cells that form strands. All cyanobacteria contain blue pigment called phycocyanin and green pigment called chorophyll *a*. However, some contain other pigments that mask the blue and green pigments. Photosynthesis is carried out in membranes that are simpler than the chloroplasts found in algae and plants.

Some students may mistake the dark areas that may appear on the prepared slides for nuclei. These areas are actually areas of concentrated chromatin. None of the organisms have nuclei.

Pre-Lab

After reading the entire procedure, ask the students to predict if they will observe distinguishable nuclei in the cyanobacteria.

Post-Lab

Have the students check their predictions based on their observations. Then ask the students how they might use the microscope to help them distinguish cyanobacteria from algae. (The cyanobacteria do not have distinguishable nuclei whereas algae do.)

INVESTIGATION 9.1

Background

Algae are photosynthetic organisms that may be unicellular or multicellular. The algae live in streams, lakes, oceans, and other bodies of water as well as in damp places on land. Most algae lack a system for moving water and nutrients from one part of the organism to another. The cells of these algae are bathed by the water in which they live. As the water passes over them, the cells obtain water and nutrients. They also eliminate wastes.

Pre-Lab

After they have read the entire investigation, ask the students to list the algae structures they expect to observe. Have them keep their lists for comparison to their actual observations.

Post-Lab

Have the students compare their lists made during the Pre-Lab activity to their observations. Encourage them to offer reasons for any differences they may have noted.

INVESTIGATION 9.2

Pre-Lab

Have the students read the entire procedure for Part A. Ask them to predict under which conditions the yeast will grow best.

Materials

Vinegar works as well as acetic acid, is inexpensive, and poses no disposal problems.

Active dry yeast is available at most grocery stores. You may wish to substitute a yeast culture from a biological supply company.

If this method of mixing the contents of test tubes is new to the students, you may wish to demonstrate the technique.

Post-Lab

Have the students review their predictions and determine if the observations supported the predictions. Then ask the students to consider if the presence or absence of light had an effect on the growth of the yeast and why or why not. (Yeast is a consumer, not a producer that needs light to help produce the nutrients needed for growth.)

Safety, Materials, and Notes

Name _____ Date _____
Class _____

CHAPTER 8 Viruses and Monerans

SECTION 1 Viruses

A. Directions: Write the correct term from the list for each description.

AIDS
immune system
nucleic acid
vaccine
host
nanometer
epidemic
polio

1. The organism invaded by a virus — **host**
2. The rapid spread of a disease through a large area — **epidemic**
3. Provides the body with the ability to fight infection — **immune system**
4. Given to people to keep them from getting a disease — **vaccine**
5. Carries the genetic, or hereditary, instructions — **nucleic acid**
6. Billionths of a meter — **nanometer**
7. A disease caused by a virus that can result in paralysis — **polio**
8. A disease caused by a virus that weakens the human immune system — **AIDS**

B. Directions: Circle the numbers of the phrases that describe characteristics of viruses.

1. (circled) Can reproduce only inside a living cell
2. Provide the body with the ability to fight infection
3. (circled) Considered hostile because they invade and then attack living cells
4. (circled) Responsible for the common cold and influenza
5. Are the material that carries hereditary instructions
6. (circled) Contain proteins and nucleic acids
7. Can prevent people from getting a disease
8. (circled) Are neither living nor nonliving

Name _____ Date _____
Class _____

CHAPTER 8 Viruses and Monerans

SECTION 2 Monerans

A. Directions: Underline the term or phrase that makes the most sense.

1. Monerans do not have (chlorophyll, <u>a nucleus,</u> a cell wall).
2. (Vaccines, Monerans, <u>Antibiotics</u>) are chemical substances that can be used to kill or slow the growth of bacteria.
3. A chemical that gives color to the tissue of living organisms is called a (<u>pigment</u>, moneran, bacterium).
4. Bacteria can cause disease by destroying cells or giving off (viruses, <u>toxins</u>, nucleic acids).
5. (Protists, <u>Bacteria</u>, Antibiotics) are necessary for the decay of waste materials.
6. (<u>Cheeses</u>, Algae, Pigments) are produced with the help of bacteria.
7. Bacteria cannot survive unless they have (a plant or an animal as a host, a cold environment, <u>enough food</u>).
8. All bacteria can (make their own food, <u>reproduce quickly</u>, cause dangerous illnesses).
9. The cyanobacteria often found in bodies of water are incorrectly called "blue-green algae" because they are *not* (monerans, <u>protists</u>, always red or brown).
10. Cyanobacteria help to restore (oxygen, carbon, <u>nitrogen</u>) to the soil.

B. Directions: Circle the numbers of the phrases that describe characteristics of monerans.

1. Cause the common cold
2. (circled) Can live alone or in groups
3. Are much larger than viruses
4. (circled) Do not have a nucleus
5. Always give off toxins
6. (circled) Always have a cell wall
7. Are always killed by antibiotics
8. Are always blue or green

8

Name _____

Class _____ Date _____

Viruses and Monerans

Chapter Review

Directions: Circle the letter of the best term for each description.

1. A very small particle that can reproduce only inside a living cell
   a. moneran   b. bacteria   **c.** virus

2. Responsible for chicken pox, measles, and AIDS
   a. bacteria   **b.** viruses   c. epidemics

3. Found in the center of a virus
   a. protein   **b.** nucleic acid   c. chlorophyll

4. Covers the center of a virus
   a. protein   b. bacteria   c. cell wall

5. Spreads rapidly among the population in a large area
   a. vaccine   b. bacteria   **c.** epidemic

6. Can prevent people from getting diseases caused by a virus
   a. vaccine   b. antibiotics   c. immune system

7. Can cure a disease caused by bacteria
   a. vaccine   **b.** antibiotics   c. epidemic

8. Chlorophyll is one example of this
   a. cyanobacteria   **b.** pigment   c. bacteria

9. Have a cell wall but do not have a nucleus
   a. viruses   b. protists   **c.** monerans

10. What a virus needs to live and reproduce
    a. host   b. oxygen   c. another virus

11. One difference between viruses and bacteria is that bacteria can
    a. live alone   b. reproduce   c. be harmful

12. Cyanobacteria always have
    a. a blue-green color   **b.** a cell wall   c. harmful effects

---

8

Name _____

Class _____ Date _____

Testing Disinfectants

(Textbook page 224) As you perform this investigation from your textbook, use this sheet to record your results and to answer the questions.

ANALYSES AND CONCLUSIONS

1. Which section(s) of the dish contained the variable(s)? The control(s)?
   Sections 1, 2, and 3 of the dish contained variables. Section 4 was the
   control.
   _____

2. Which disinfectant appears to be the most effective? How can you
   tell?
   The paper disc with the largest clear area (area of inhibition) around it
   had the most effective disinfectant.
   _____

3. The agar and the Petri dish were sterilized before this investigation.
   Why do you think this was necessary?
   Bacteria were already present on the unsterilized agar and Petri dish. If
   these bacteria were not destroyed by sterilization, they would grow on
   the agar plates, and the results from the swabbing would be
   inaccurate.
   _____

4. What is the purpose of a control in an investigation?
   A control is used to compare the effects of a variable against a stable
   condition. The control of an experiment reveals what would have
   happened if nothing were changed.
   _____

(continues)

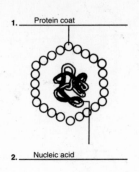

1. Protein coat

2. Nucleic acid

3. Nucleic acid not visible

4. Protein coat

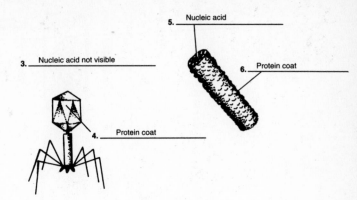

5. Nucleic acid

6. Protein coat

1. How are viruses similar to living things? How are viruses different from living things?

2. Describe the shapes of viruses.

Name _____ Class _____ Date _____

INVESTIGATION 8

Testing Disinfectants (continued)

APPLICATION

Why do you think it is important to wash your hands thoroughly at the end of this investigation? List several ways in which this procedure applies to your everyday life.

It is important to wash your hands to make sure that they are free of

bacteria. Because many bacteria can be harmful to people, hands should be

washed frequently to prevent the spread of bacteria. It is especially

important to wash your hands before eating.

_____

_____

_____

_____

21

Reproduction of Viruses

HOLT LIFE SCIENCE
8-3
Page 212

1. Virus attaches to cell. Viral nucleic acid enters cell.

— Nucleic acid

2. Viral nucleic acid replicates.

New viruses

4. Cell bursts releasing new viruses.

3. Cell makes new viruses.

1. Why are viruses considered to be hostile to living things?

2. Describe a situation, if one exists, in which a cell would be a willing host to a virus. Explain your reasoning.

Name
Class

Date

INVESTIGATION

8.1

Making Models of Viruses

Purpose

■ To use information from a chart to make accurate scale models of several viruses

Materials

Metric ruler
Plastic foam
Clay
Insulated electrical wire
Cork

Toothpicks
Pipe cleaners
Wooden dowels
Tape
Construction paper

Procedure

1. Look at Table A. The illustrations in the first column show the basic shape of each virus. However, the drawings are not drawn to scale. That is, the real difference in the sizes of the viruses is not shown. Read the third column of Table A to find the actual size of each virus. Sizes are given in nanometers (nm). One nanometer is equal to 0.000 000 001 m. NOTE: Length values are not given for viruses that are more or less spherical in shape.

TABLE A: SIZES OF VIRUSES

Shape of Virus	Type of Virus	Actual size Diameter in nm	Actual size Length in nm	Relative size Diameter	Relative size Length	Scale size Diameter in mm	Scale size Length in mm
	Potato X	10	500	1.0	50	5	250
	Polio	28		2.8		14	300

(continues)

INVESTIGATION 8.1   Making Models of Viruses (continued)

TABLE A: SIZES OF VIRUSES (continued)

Shape of Virus	Type of Virus	Actual size Diameter in nm	Actual size Length in nm	Relative size Diameter	Relative size Length	Scale size Diameter in mm	Scale size Length in mm
	Influenza	100		10.0		50	200
	Mumps	200		20.0		100	100
	Tobacco Mosaic	18	300	1.8	30	9	150

2. The easiest way to find the relative sizes of the viruses is to assign them numbers that are easier to work with. The fourth column of Table A shows that the diameter of the Potato X virus has been assigned a relative size of 1. Notice that relative measurements have no units. This is the smallest virus in the table. To give the diameter of this virus a size of 1, you must divide its actual diameter by 10 nm:

$$10 \text{ nm} \div 10 \text{ nm} = 1.$$

3. Find the relative length of the Potato X virus by dividing its actual length by 10 nm. Record your result in the appropriate place in the fourth column.

4. To keep the relative sizes of the viruses the same, you must now perform the same procedure for each of the other viruses. Divide each diameter and length by 10 nm. Record your results in the appropriate places in the fourth column.

5. Before you can make your models of the viruses, you must decide on a scale size. You could use the relative sizes that you have already found. However, if you used the relative sizes as they are, some of your models might become too large to handle easily. You need to use a scale on which the smallest virus is easy to make, but on which the largest virus is not too large. For convenience, assume that the relative size of 1 equals 5 mm. To do this, multiply the relative diameter (1) by 5 mm. Then, do the same for the relative length: 50 × 5 mm = 250 mm. This gives the Potato X virus a scale size of 5 mm × 250 mm.

(continues)

INVESTIGATION 8.1   Making Models of Viruses (continued)

6. Follow the same procedure to find the scale sizes of the other viruses. Make one scale model for each virus. Be careful to follow the scale sizes you determined in item 6.

7. Choose the best materials to use to make each virus. Make one scale model for each virus. Be careful to follow the scale sizes you determined in item 6.

Analyses and Conclusions

1. How can models such as those you have created help people understand the relative sizes of viruses?

    The students may say that seeing three dimensional models can help

    people understand what the dimensions (numbers) mean.

2. What conclusion can you make about the relative sizes of the mumps virus and the influenza virus?

    The mumps virus is twice the size of the influenza virus.

3. Can you make a valid comparison of the size of the influenza virus and the Potato X virus? Explain your answer.

    The students should understand that it is difficult to compare the sizes

    of organisms whose shapes are so different.

Application

1. Review the information in your textbook regarding the way viruses enter cells and reproduce. In the space provided, write a paragraph explaining how the various shapes of viruses may help the viruses gain entry into cells and do their work.

    Accept any responses that the students can support with scientific

    evidence or logic.

(continues)

INVESTIGATION 8.1 Making Models of Viruses (continued)

2. Write a paragraph discussing the variety of sizes in viruses. In your paragraph, suggest possible answers to the following questions:

a. What relationship might the size of a virus have to the type of cell it invades?

_____
_____
_____
_____
_____
_____
_____
_____
_____
_____
_____

b. Is there such a thing as a typical virus? Explain your answer.

Accept any responses that the students can support with scientific evidence or logic.

_____
_____
_____
_____
_____
_____
_____
_____
_____

INVESTIGATION 8.2 Observing Cyanobacteria

Purpose

■ To study the characteristics of several types of cyanobacteria

Materials

Prepared slides of
  Gloeocapsa, Anabaena,
  Oscillatoria, and *Nostoc*
Compound light microscope

Procedure

1. Cyanobacteria are a group of organisms similar to many other bacteria. However, cyanobacteria contain chlorophyll, just as is found in green plants. Place a prepared slide of *Gloeocapsa* under a microscope. Observe the slide under low and high power. This type of cyanobacteria is unicellular. Note the circles that seem to surround each cell. These are actually layers of jellylike material. In the space provided, sketch the high-power view of the *Gloeocapsa.* Then, describe any distinguishing characteristics on the lines beside your sketch.

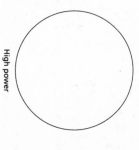

High power

The students should note the single cells and the sheaths that surround them.

_____
_____
_____

(continues)

INVESTIGATION 8.2 Observing Cyanobacteria (continued)

[High power]

The students should note that
Nostoc grows in groups or
clusters of filaments.

Analyses and Conclusions

1. How is *Gloeocapsa* different from the other cyanobacteria?
It is unicellular.

2. What differences did you observe among the slides of *Nostoc*, *Oscillatoria*, and *Anabaena*?
The students may notice that *Anabaena* forms single filaments.
Oscillatoria and *Nostoc* both form colonies of filaments.

Application

The growth rate of most kinds of cyanobacteria is directly related to the chemical content of the water in which it grows. Use reference sources to find out what **algal bloom** is. Write a short essay explaining what algal bloom is, its effect on the environment, and how it can be prevented.

The students should be able to identify algal bloom as the green covering on some lakes and ponds that results from the uncontrolled growth of cyanobacteria and some types of algae. Algal bloom decreases the amount of O_2 in the water, causing fish to die and making the water unfit for human or animal consumption. It is caused by a high content of phosphates and some other chemicals in the water. It can be prevented by keeping sewage containing these chemicals from being dumped into the water.

INVESTIGATION 8.2 Observing Cyanobacteria (continued)

2. Place the slide of *Anabaena* under the microscope and observe it under low and high power. *Anabaena* is a multicellular cyanobacteria that forms long filaments, which are one cell thick. In the space provided, sketch a high-power view of *Anabaena*. Then, describe any distinguishing characteristics on the lines beside your sketch.

[High power]

The students should notice that
Anabaena grows in single
filaments.

3. Place the slide of *Oscillatoria* under the microscope and observe it under low and high power. *Oscillatoria* is another cyanobacteria that forms long filaments. However, it is different in many respects from the *Anabaena* you studied in Step 2. In the space provided, sketch the high-power view of the *Oscillatoria*. Then, describe its characteristics on the lines beside your sketch. Be sure to include characteristics that distinguish *Oscillatoria* from *Anabaena*.

The students should note that
Oscillatoria grows in clusters of
filaments.

4. Place the slide of *Nostoc* under the microscope and observe it under low and high power. *Nostoc*, too, forms long, thin filaments. The arrangement of the filaments, however, differs from both *Anabaena* and *Oscillatoria*. In the space provided, sketch the high-power view of *Nostoc*. Then, describe the characteristics of *Nostoc* on the lines beside your sketch. Be sure to include characteristics that distinguish *Nostoc* from *Anabaena* and *Oscillatoria*.

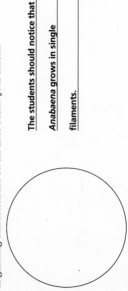

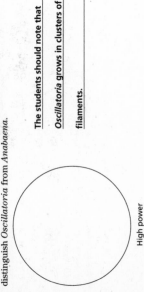

[High power]

(continues)

CHAPTER 8 Viruses and Monerans

Reading Skills

Making a Chapter Outline

Directions: Read Chapter 8. Then, on a separate sheet of paper, write a complex outline of the information in the chapter. A complex outline is one in which you include minor details as well as major ones. The format for a complex outline is shown below. Be sure to follow this format when you complete your outline.

Title of Outline

I. Main Idea 1
  A. Supporting detail for main idea 1
  B. Supporting detail for main idea 1
    1. Supporting detail for detail B
      a. Supporting detail for detail 1
      b. Supporting detail for detail 1
    2. Supporting detail for detail B
    3. Supporting detail for detail B
  C. Supporting detail for main idea 1
II. Main Idea 2
  A. Supporting detail for main idea 2
    1. Supporting detail for detail A
    2. Supporting detail for detail A
    3. Supporting detail for detail A
  B. Supporting detail for main idea 2
III. Main Idea 3
  A. Supporting detail for main idea 3—one supporting detail for detail A
  B. Supporting detail for main idea 3
    1. Supporting detail for detail B
    2. Supporting detail for detail B

Notice that when a main idea or supporting detail is modified by only one detail, the detail is written on the same line as the idea it modifies (see III. A).

Outlines should include a logical organization of the information in the textbook.

HRW material copyrighted under notice appearing earlier in this work.

17

CHAPTER 8 Viruses and Monerans

Science and Social Studies

Modern Medicine and Society

Infectious diseases have strongly influenced the course of civilization throughout human history. The Black Plague, for example, brought about the collapse of the feudal system of medieval Europe by killing off the majority of the working class. Major battles have been decided because armies were conquered by diseases, not by enemies. Infectious diseases have long dictated the areas in which people could settle. Even social contacts within a community can be affected by outbreaks of infectious diseases. The majority of the infectious diseases caused by bacteria can now be controlled through the use of antibiotics. Altered toxins, weakened live viruses, and dead organisms are all used today to immunize people against many life-threatening viral diseases.

A. Directions: Do research to find out how the control of diseases by using antibiotics and immunizations affects society. For example, when life-threatening diseases are controlled, people tend to live longer. How does this affect society in general? Then, on a separate sheet of paper, write an essay presenting the information you find.

Essays will vary, but the students should understand that when life-threatening diseases are controlled and people live longer, society may be faced with problems such as caring for older people, providing for people who are unable to work, and so on.

B. Directions: In the space provided, answer the following questions.

1. List two infectious diseases that still cannot be controlled by antibiotics or immunization.

The students may list the common cold and AIDS among others.

2. Why are the leading causes of death no longer the infectious diseases that killed so many people at the beginning of this century?

because those diseases have now been controlled by antibiotics and immunization

3. Scientists may someday be able to conquer all types of disease. What effect do you think this would have on society? Explain your answer.

The students may say that the eradication of all disease might be harmful to society because the earth might become overpopulated, creating housing problems and food shortages; or they may say that society would benefit because the dollars spent each year on disease research could be applied to solving food shortages.

18

HRW material copyrighted under notice appearing earlier in this work.

84

C H A P T E R
8

Viruses and Monerans

Extending Science Concepts

Calculating Bacterial Growth

Bacteria reproduce very rapidly. One bacterium can reproduce in about 20 minutes if conditions are favorable. After a bacterium reaches full size, it divides into two. The two bacteria then grow to full size and each divides into two again, forming four bacteria. If conditions are favorable, one bacterium can form millions of bacteria in a few short days.

To see how many bacteria can be formed in just six hours, design a table similar to Table A. The table should show for every 20-minute interval, the number of bacteria formed. Start with one bacterium at time 0:00.

TABLE A: BACTERIAL GROWTH

Time	Number of bacteria	Time	Number of bacteria
0:20	2	3:20	1024
0:40	4	3:40	2048
1:00	8	4:00	4096
1:20	16	4:20	8192
1:40	32	4:40	16 384
2:00	64	5:00	32 768
2:20	128	5:20	65 536
2:40	256	5:40	131 072
3:00	512	6:00	262 144

(continues)

HRW material copyrighted under notice appearing earlier in this work.

19

CHAPTER 8

Viruses and Monerans *(continued)*

Make a graph of the six hours of bacterial growth shown in Table A. Then answer the following questions.

1. How many cell divisions take place in one hour? In five hours?

   3, 15

2. Why do you think bacteria seldom ever multiply as rapidly as the table shows?

   Students might suggest that space and food are limited and too much

   waste would accumulate.

3. Explain the following statement: Bacteria multiply by dividing.

   A dividing bacterium produces 2 new bacteria.

20

HRW material copyrighted under notice appearing earlier in this work.

HRW material copyrighted under notice appearing earlier in this work.

85

CHAPTER 8
Viruses and Monerans

Thinking Critically

Directions: Read about an experiment that Dr. Wilson performed. Then answer the questions about her experiment.

Dr. Erica Wilson has found a strain, or type, of bacteria called *Staphylococcus aureus* that she believes is causing a widespread infection among her patients. Dr. Wilson knows that there are many different strains of *S. aureus*. An antibiotic that kills one strain of the bacteria may not kill another strain. Therefore, Dr. Wilson set up an experiment to see which antibiotic she should use to fight the infection. This is the procedure she followed.

1. She obtained a pure sample of the strain of *S. aureus* she wanted to test.

2. She obtained a sterile Petri dish that contained **agar**, a jelled substance that provides food for the bacteria. The agar contained the same food sources that are available to bacteria inside the human body.

3. She spread a tiny amount of the bacteria onto the agar, being careful not to contaminate her sample or the agar.

4. She placed five antibiotic test discs onto the agar on top of the bacteria. The test discs she used were about 3 mm in diameter. Each test disc contained a different antibiotic in the amount usually given to people to fight infections.

5. On the bottom of the dish, she numbered each disc so she could tell later which one was which. (See Fig. A.) She made the following notes in her notebook:

Figure A
1 = Penicillin
2 = Streptomycin
3 = Tetracycline
4 = Ampicillin
5 = Sulfanilamide

6. She covered the dish to avoid contamination. Then she placed it in an incubator (an ovenlike instrument that keeps a constant temperature). She set the incubator at 37°C (human body temperature).

(continues)

CHAPTER 8 Thinking Critically *(continued)*

7. After 48 hours, she looked at the dish. This is what she saw:

1. What might account for the fact that the bacteria did not grow around some of the test discs?
Students should say that some of the antibiotics were effective in fighting growth of this strain of *Staphylococcus aureus*.

2. Why did the bacteria grow closer to some test discs than others?
Some antibiotics are more effective than others against this particular strain of *S. aureus*.

3. Why was it necessary to make sure that the agar and the sample were not contaminated?
The students may say that contamination may have caused false results or that the experiment may not have worked.

4. Ampicillin is derived from penicillin. Why do you think these two antibiotics showed different amounts of "clear zone"?
Students may say that the two antibiotics have different properties or that *S. aureus* is more susceptible to one than the other.

5. Judging from the results of the experiment, which antibiotic should Dr. Wilson give to patients to fight the infection?
Sulfanilamide. Note to the Teacher: Normally, other factors would also be taken into consideration, such as the patients' allergies to certain antibiotics.

CHAPTER TEST Viruses and Monerans *(continued)*

c **9.** Which of these activities could result in the transmission of the AIDS virus?
   a. shaking hands with an AIDS-infected person
   b. breathing the same air as an AIDS-infected person
   c. having sexual contact with an AIDS-infected person
   d. having an AIDS-infected person as your friend

a **10.** Which of these diseases is caused by bacteria?
   a. strep throat   b. rabies
   c. measles   d. influenza

d **11.** Choose the statement that is most correct.
   a. Bacteria and viruses exist in identical conditions.
   b. Bacteria and viruses exist in similar conditions.
   c. Viruses can exist in conditions that bacteria cannot exist in.
   d. Bacteria can exist in conditions that viruses cannot exist in.

b **12.** Cyanobacteria are an example of
   a. viruses.   b. monerans.
   c. pigments.   d. antibiotics.

Interpreting Graphics

13. These drawings show three typical bacteria. Describe the shapes of typical bacteria.

Most bacteria are shaped like cylinders or rods, spirals, or spheres.

Reviewing Themes

14. Systems and Structures
Did scientists and researchers know that microscopic bacteria and viruses existed before the invention of microscopes? Explain.

No. They could only guess what caused the diseases they saw.

(continues)

CHAPTER TEST 8 Viruses and Monerans

Understanding Vocabulary

Explain how the terms in each pair are related.

1. host, nucleic acid of a virus
The nucleic acid of a virus controls the activities of a host cell.

2. AIDS, immune system
The AIDS virus disables the immune system of a person, making that person extremely vulnerable to infection.

3. antibiotic, vaccine
Antibiotics and vaccines work to slow or prevent the spread of a disease.

4. virus, bacteria
Microscopic viruses and bacteria affect living things.

5. epidemic, time
An epidemic is the spread of a disease through a large area in a relatively short time.

6. bacteria, toxins
Some bacteria release toxins that damage living things.

7. pigments, chlorophyll
Chlorophyll is the pigment, or chemical, that makes plants green.

Understanding Concepts

MULTIPLE CHOICE
In the space to the left, write the letter of the choice that best completes the statement or answers the question.

d **8.** Viruses are able to reproduce
   a. anywhere in the environment of Earth.
   b. inside the living cells of plants.
   c. inside the living cells of animals.
   d. inside the living cells of plants and animals.

(continues)

CHAPTER TEST Viruses and Monerans (continued)

15. *Environmental Interactions*
What would happen to plants if cyanobacteria became extinct? **Cyanobacteria restore nitrogen to the soil. Since nitrogen is a nutrient necessary for proper plant growth, plant growth would be negatively affected by a lack of nitrogen.**

Thinking Critically

16. The shapes of human beings are similar in the sense that most have two legs, a head, and a trunk. Are the shapes of viruses also similar? **Viruses have many different shapes and sizes.**

17. Must one or more people die before a disease can be called an epidemic? Explain. **Epidemics do not always result in fatalities. For example, a rapid spread of influenza through a large area may affect many people, but the people recover in time. Fatalities are, however, likely during the course of many epidemics.**

18. Bacteria reproduce very quickly. Why aren't more bacteria found on Earth? **The conditions needed by bacteria to reproduce are seldom perfect and often isolated.**

19. How can a person lessen the likelihood of acquiring the AIDS virus? **Avoiding the activities that are known to transmit the AIDS virus will significantly lessen the likelihood of acquiring the virus.**

20. Knowing what you know about viruses, is it possible for a person to contract a cold from exposure to cold weather? **The common cold is caused by the spread of a virus, not by exposure to cold weather.**

Performance Assessment
Observing Bacteria

Bacteria can be found almost everywhere in the world around you. However, many people have never seen what different bacteria look like. This investigation will allow you to observe various bacteria.

To complete this task in one class period, the students should work in groups of 3 or 4 students. The students will need the following materials: 100-mL beaker, tap water, medicine dropper, microscope slide, coverslip, plain yogurt or sour cream, methylene blue, and compound light microscope.

CHAPTER TEST Viruses and Monerans (continued)

THE MATERIALS

100-mL beaker • tap water • medicine dropper • microscope slide • cover-slip • plain yogurt or sour cream • methylene blue • compound light microscope

THE INVESTIGATION

Part A. Readying the materials
Fill a beaker with approximately 25 mL of water. Then add a small amount of plain yogurt or sour cream to the water and stir thoroughly. Using a medicine dropper, place a drop of the beaker mixture on a slide and stain it using a drop of methylene blue. Cover the slide with a coverslip and mount it on a microscope. Remember to follow the proper procedure when using a compound light microscope. **The students should correctly prepare a slide and mount that slide in the staging area of each microscope.**

Part B. Making your observations
Using the different objectives of your microscope, locate bacteria. **The students should follow a procedure that does not damage a slide when switching from one objective to another.**

Part C. Gathering your data
Note the general shape and type of movement that the bacteria display. Then create a sketch of the different bacteria you observe, and include with each sketch a brief note describing the movement of the bacteria.
The sketches created by students should be reasonably accurate renditions of what they observe and should contain a brief description of the movement (if any) of the observed bacteria.

Part D. Going beyond your investigation
Recall that bacteria can be found almost everywhere in the world around you. Think of other substances or locations that you suspect may contain bacteria; then design and perform an experiment that will test for the presence of those bacteria. **The students should design an experiment that will test for the presence of bacteria.**

Name _____
Class _____ Date _____

CHAPTER 9
Protists and Fungi

SECTION 1 Protozoans

A. Directions: In the space provided, write the letter of the term in the box that best fits the description.

e **1.** In amoebas, these structures collect extra water from the cytoplasm and release it through the cell membrane.

d **2.** These long, whiplike structures are used for movement in some protozoans.

h **3.** These move by using long, whiplike structures.

c **4.** These hairlike structures cover the entire cell and can create currents of water to bring food to the organism.

i **5.** All of these are parasites. They cannot move from place to place on their own.

g **6.** These are the most complex protozoans. They are covered with hundreds of hairlike structures.

b **7.** These projections made of cytoplasm are used for movement and to capture food.

f **8.** These protozoans move from place to place using pseudopods.

a **9.** This microscopic organism is single-celled and is most likely to be found in water. One kind causes malaria.

a. protozoan
b. pseudopods
c. cilia
d. flagella
e. contractile vacuole
f. sarcodines
g. ciliates
h. flagellates
i. sporozoans

B. Directions: Match each organism in Column A with the group of protozoans in Column B to which it belongs.

COLUMN A

c **1.** amoeba

a **2.** paramecium

b **3.** parasite that causes African sleeping sickness

d **4.** parasite that causes malaria

COLUMN B

a. ciliates
b. flagellates
c. sarcodines
d. sporozoans

Name _____
Class _____ Date _____

CHAPTER 9
Protists and Fungi

SECTION 2 Algae

Directions: Write the correct term from the list below to complete each sentence.

dinoflagellates	red tide	shells	*Euglena*
diatomaceous earth	diatoms	algae	fire algae
photosynthesis	pigments		

1. Often found in water and sometimes on tree trunks, __algae__ are members of the protist kingdom that have chloroplasts in their cells.

2. Algae use a process called __photosynthesis__ to make their own food from sunlight and nutrients.

3. Algae are classified according to the red, green, or brown __pigments__ they contain.

4. One type of freshwater algae is the __Euglena__, an organism that has characteristics of both plants and animals.

5. The __dinoflagellates__ are a type of algae found in the ocean. Each organism moves using two flagella.

6. When dinoflagellates reproduce very rapidly and release poisons into ocean water, they can cause a __red tide__ that kills fish.

7. Some dinoflagellates are nicknamed __fire algae__ because they glow like tiny fireflies when they are disturbed.

8. The __diatoms__ are golden brown algae that are the most common of all the single-celled organisms in the oceans.

9. The cell walls, or __shells__, of diatoms are in two parts and made of a chemical similar to glass.

10. The empty shells of diatoms collect on the ocean floor to form __diatomaceous earth__, a material used in many products.

SECTION 3   Fungi

A. Directions: Write one term from the list below for each description. Write each letter of the term on a separate line or box.

| slime molds | spores | lichen |
| hyphae | fungi | mycelium |

1. Organisms with bodies made up of hyphae that decompose organic material

   f __ u __ n __ g __ i __

2. Part fungus and part alga

   __ __ __ __ __ __ __
   l   i   c   h   e   n

3. Move like amoebas, look and reproduce like fungi

   s __ l __ i __ m __ e __ __ __ __ __ __ __
                                   m o l d s

4. A group of interlocking hyphae

   m __ y __ c __ e __ l __ i __ u __ m __

5. Threadlike structures that make up the bodies of fungi

   h __ y __ p __ h __ a __ e __

6. Formed by most fungi in order to reproduce

   s __ p __ o __ r __ e __ s __

B. Directions: Unscramble the letters in the boxes to find the term for one of the most helpful members of the fungus kingdom.

penicillin _____

HRW material copyrighted under notice appearing earlier in this work.

29

Chapter Review

Directions: Circle the letter of the best term for each description.

1. The protozoans are classified into four groups. Which of these is not one of those four groups?
   a. ciliates   b. parasites   c. sporozoans

2. Amoebas use these for movement and to capture food.
   a. pseudopods   b. cilia   c. flagella

3. Amoebas use contractile vacuoles for this function.
   a. store food   b. collect water   c. store waste

4. African sleeping sickness is caused by these organisms.
   a. tsetse flies   b. flagellates   c. flagella

5. Algae are important in ocean food chains because they can do this.
   a. make food   b. reproduce   c. decompose waste

6. These chemicals are used to classify algae.
   a. chlorophyll   b. chloroplasts   c. pigments

7. Which of these algae have two-part shells?
   a. dinoflagellates   b. Euglena   c. diatoms

8. Which of these algae live in fresh water?
   a. Euglena   b. diatoms   c. dinoflagellates

9. These organisms decompose organic material.
   a. algae   b. parasites   c. fungi

10. These organisms obtain food from dead organisms or waste products.
    a. parasites   b. saprophytes   c. algae

11. Which of these reproduce by forming spores?
    a. most fungi   b. most algae   c. protozoans

12. Lichens are not classified as fungi because they can do this.
    a. retain water   b. make food   c. move to a new place

30

HRW material copyrighted under notice appearing earlier in this work.

90

HRW material copyrighted under notice appearing earlier in this work.

Name _____ Class _____ Date _____

INVESTIGATION 9

9 | Observing Protozoans

(*Textbook page 235*) As you perform this investigation from your textbook, use this sheet to record your results and to answer the questions.

PROCEDURE

4. Using low power on your microscope, observe the drop of water. Draw and label what you see.

5. Switch to high power. Again draw and label what you see.

TABLE 1: OBSERVATION OF PROTOZOANS

Protist Name	Sketch	Description (Size/Shape/Color)	Type of Movement	Type of Feeding

(*continues*)

31

Name _____ Class _____ Date _____

INVESTIGATION 9   Observing Protozoans (*continued*)

ANALYSES AND CONCLUSIONS

1. What types of protozoans do you see? Flagellates? Ciliates? Amoebas? Ciliates are the protozoans most often observed. Some flagellates may also be seen. Amoebas may be observed if pond water is used along with some debris from the bottom of the container.

2. Compare your drawings with those of your classmates. How are the organisms in your samples similar to and different from the organisms in their drawings?

   Drawings should include numerous ciliates and may include flagellates and amoebas. If the larvae of small crustaceans appear, have the students note their jointed appendages and multicellular nature. These structures will help the students distinguish crustaceans from protozoans.

3. How do the organisms from the different water samples compare? The organisms in the water samples will depend on the source of the water. Pond water from a pond's edge is likely to show a variety of organisms.

APPLICATION

Why is it important for a scientist or a technician working in a water purification plant to be able to identify protozoans? Explain your reasoning. Protozoans can cause human diseases. Also protozoans in a water supply may indicate the unwanted presence of organic materials.

32

22

Protists

HOLT LIFE SCIENCE
9-3, 9-6
Pages 231, 233

1. Cell membrane _____
2. Cytoplasm _____
3. Nucleus _____
4. Pseudopods _____
5. Food vacuoles _____
6. Contractile vacuole _____

Amoeba

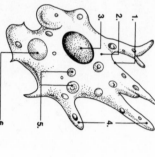

Paramecium

7. Contractile vacuole (full) _____
8. Food vacuole _____
9. Cytoplasm _____
10. Nucleus _____
11. Oral groove _____
12. Cilia _____
13. Anal pore _____
14. Contractile vacuole (empty) _____

1. Why is it difficult to describe the shape of an amoeba?
2. Describe how an amoeba moves from place to place.
3. How do the cilia of a paramecium facilitate movement from place to place?

23

Fungi

HOLT LIFE SCIENCE
9-14, 9-15
Pages 241, 242

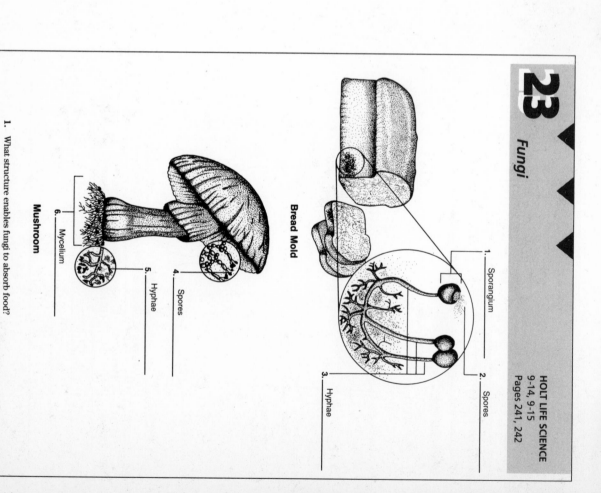

Bread Mold

1. Sporangium _____
2. Spores _____
3. Hyphae _____

Mushroom

4. Spores _____
5. Hyphae _____
6. Mycelium _____

1. What structure enables fungi to absorb food?
2. Is a living organism ever a willing host to fungi? Why or why not?
3. How do most fungi reproduce?

INVESTIGATION 9.1 — Observing and Classifying Algae (continued)

TABLE A: CHARACTERISTICS OF ALGAE

Type of Algae	Shape of Algae	Unicellular or Multicellular	Flagella Present?	Other Characteristics
Chlamydomonas	egg-shaped	unicellular	yes	reddish pigment (stigma) near flagellar end
Scenedesmus	rectangular	multicellular	no	has spines; forms colony by arranging long cells side by side
Spirogyra	long filaments	multicellular	no	spiral chloroplasts; cells attached end to end
Volvox	rounded	multicellular	yes	forms spherical colony; has specialized cells for reproduction

Analyses and Conclusions

1. Which of the four types of algae do you think are able to move around freely? Explain your answer.

   Chlamydomonas and *Volvox*, because they have flagella with which to

   propel themselves

2. Of the multicellular algae that you examined, which do you think are the most likely to have specialized cells?

   Volvox has specialized cells for reproduction. The students should be

   able to see some of these cells on the prepared slide. *Spirogyra* and

   Scenedesmus do not contain specialized cells.

Application

All of the algae you studied in this investigation are freshwater algae. *Spirogyra* and *Chlamydomonas* live primarily in polluted water. How can knowing about these organisms help your community?

Answers should reflect the knowledge that particular types of algae can

indicate the presence of pollution in bodies of water.

INVESTIGATION 9.1 — Observing and Classifying Algae

Purpose
- To examine the characteristics of several different kinds of algae

Materials
Compound light microscope
Prepared slides of
Chlamydomonas, Scenedesmus, Spirogyra, and *Volvox*

Procedure

1. Place a prepared slide of *Chlamydomonas* under the microscope. Examine the algae under low and high power. Sketch the organism.

2. Record the characteristics of *Chlamydomonas* in Table A. In the column labeled "Other Characteristics," list any characteristics you find interesting or that you think will serve to distinguish this organism from other algae.

3. Repeat steps 1 and 2 for *Scenedesmus, Spirogyra,* and *Volvox.* Be sure to label your sketches.

(continues)

INVESTIGATION 9.2

Observing the Growth of Yeast

Purpose
- To determine the best growing conditions for yeast
- To observe the effects of pollution on the growth of yeast

Materials

Wax pencil
Test tubes (12)
Forceps
Active dry yeast
Granulated table sugar
Graduate, 15 mL

Distilled water
Test-tube racks (2)
Aluminum foil
Medicine dropper
Liquid dish detergent
Acetic acid solution, 5 percent

Procedure

Part A: Determining the best growing conditions for yeast

1. Using a wax pencil, label 8 test tubes with the numbers 1 through 8.
2. Using forceps, transfer 2 grains of dry active yeast into test tubes 3, 4, 7, and 8. Then transfer 4 granules of sugar to test tubes 5, 6, 7, and 8.
3. Add 5 mL of distilled water to all 8 test tubes. Roll each test tube gently between your hands to mix the contents.
4. Use Table A to check your test-tube setup.

TABLE A: TEST-TUBE SETUP

Test Tube	Distilled Water	Dry Yeast	Sucrose	Light	Dark
1	×			×	
2	×				×
3	×	×		×	
4	×	×			×
5	×		×	×	
6	×		×		×
7	×	×	×	×	
8	×	×	×		×

(continues)

INVESTIGATION 9.2

Observing the Growth of Yeast (continued)

5. Place all the even-numbered test tubes in one test-tube rack, and place all the odd-numbered test tubes in the other rack.
6. Wrap the even-numbered test tubes loosely in foil to keep them in the dark, or store them in a dark cabinet. Leave the odd-numbered test tubes in a lighted area.
7. In Table B, note the appearance of the contents of test tubes.

TABLE B: APPEARANCE OF TEST TUBES

Test Tube Set	Beginning Appearance	Appearance after 24 Hours
1	clear	clear
2	clear	clear
3	slightly opaque	slightly opaque
4	slightly opaque	slightly opaque
5	clear	clear
6	clear	clear
7	slightly opaque	very cloudy
8	slightly opaque	very cloudy

8. After 24 hours check the test tubes. In Table B, note the appearance of the contents of the test tubes.

Part B: Observing the effects of pollution on the growth of yeast

1. Label four test tubes A, B, C, and D. Determine which environment in Part A is best suited for growing yeast. Add those ingredients to all four test tubes. Test tube A will be a control.
2. Add two drops of liquid dishwashing detergent to test tube B.
3. Add two drops of 5 percent acetic acid to test tube C.
4. Add two drops of liquid dishwashing detergent and two drops of 5 percent acetic acid to test tube D. Fill in Table C to describe the contents of your test tubes.
5. Roll each tube gently between your hands to mix the contents.

(continues)

INVESTIGATION 9.2   Observing the Growth of Yeast (continued)

TABLE C: TEST-TUBE CONTENTS

Test Tube	Distilled Water	Dry Yeast	Sucrose	Detergent	Acetic Acid
A	x	x	x		
B	x	x	x	x	
C	x	x	x		x
D	x	x	x	x	x

6. In Table D, note the appearance of each test tube.

TABLE D: APPEARANCE OF TEST TUBES

Test Tube	Beginning Appearance	Appearance after 24 Hours
A	slightly cloudy	cloudy
B	slightly cloudy, has suds	almost clear
C	slightly cloudy	almost clear
D	slightly cloudy, has suds	slightly cloudy

7. Allow the test tubes to stand for 24 hours in a light or dark environment, depending on your results in Part A.

8. After 24 hours, check the test tubes. In Table D, note the appearance of each set of test tubes.

Analyses and Conclusions

1. According to your results in Part A, what is the best environment for growing yeast?

   Answers should note that the presence of sucrose seems to be

   necessary for the rapid growth of yeast.

(continues)

41

INVESTIGATION 9.2   Observing the Growth of Yeast (continued)

2. After performing the investigation in Part A, can you say that you have found the one best environment for growing yeast? Explain your answer.

   The students should realize that many variables have not yet been

   tested—for example, the temperature of the water or growth in a solid

   or semisolid medium. They should also realize that there is more than

   one kind of yeast.

3. In Part A, why do you think it was necessary to have one set of test tubes that contained nothing but water?

   The students may say that the water sets were used as controls to

   make sure nothing in the water itself would make the water cloudy in

   24 hours.

4. Look at your results for Part B. What effect did the detergent have on the growth of yeast?

   The students may say that the yeast did not grow well in the presence

   of the detergent.

5. What effect did the acetic acid have on the growth of yeast?

   The growth of the yeast should be significantly slower.

Application

Adding pollutants such as detergent and acid to a yeast culture made the yeast behave differently than it normally would. Explain what might happen to the water in lakes and rivers if large amounts of these pollutants were allowed to enter the water.

Answers should reflect an understanding that the pollutants would

seriously affect, if not destroy, the life in the body of water.

42

CHAPTER 9 Protists and Fungi

Reading Skills

Taking Notes

When you read a chapter, it sometimes helps to take notes about what you are reading. The procedure for taking notes is similar to that for making an outline. However, the procedure for taking notes is much less structured.

To take notes, first be sure you have the right equipment. You will need a pen or several sharpened pencils and some clean sheets of paper or a notebook. Then, as you read, write down ideas that you want to remember. These will usually include the main ideas and the major supporting details of a chapter.

Many people associate note taking with quickly scribbled, abbreviated notes that are almost unreadable. Although this is the way many people take notes, it is not the correct way. Effective notes must be easy to understand. There are many things you can do to make your notes easy to understand. Use the following guidelines to write notes that are good enough to use later to study for tests.

■ At the top of the page, write the name and number of the chapter about which you are taking notes. This will help you identify the notes easily later.

■ Always write clearly and neatly. Notes will do you little good if you cannot read them later.

■ You may use abbreviations if you are sure that their meanings are absolutely clear to you. You must be able to go back to them later and know exactly what they mean. For example, the abbreviation *sub.* might stand for *substance, subject, subtopic,* or *substitute.* Abbreviations that can stand for more than one word may only confuse you later.

■ Sometimes it helps to add a sentence or two to explain an idea more thoroughly. However, it is not necessary to write down every minor detail. Keep your notes as short and to the point as you can.

■ After you have taken notes on the whole chapter, go back and review your notes immediately. If they seem clear, you are finished. However, if they seem disorganized, you may wish to rewrite your notes in what seems to you a more logical order. Add any details that are necessary to understand the notes.

(continues)

CHAPTER 9 Reading Skills (continued)

■ After you have finalized your notes, keep them together in a notebook or other place where you can find them easily.

Directions: Reread Chapter 9 of your textbook. In the space provided, write notes about the chapter. Be sure to follow the guidelines discussed. If you think you need to rewrite or reorganize your notes, do so on another sheet of paper and attach it to this page.

The students' notes should include information about the characteristics

of protists and the different types of protists and their structures. In

taking notes on the algae, the students should identify the characteristics

of algae as well as note that most scientists classify algae as protists.

Notes on the fungi should identify the different types of fungi and their

characteristics. The students should also note the characteristics of slime

molds and lichens.

Science and Language

Writing a Biographical Report

Dr. Alexander Fleming was born in Scotland and educated in London. He served as a professor of bacteriology from 1928 to 1948 at St. Mary's Hospital Medical School of London University. Dr. Fleming is recognized as an outstanding researcher in the fields of bacteriology, immunology, and chemotherapy. However, he is best known for a discovery he made accidentally in 1928, while working with a culture of staphylococcus [staf uh loh KAHK uhs], a type of bacteria. He noticed that a penicillium mold had stopped the growth of the bacteria. His discovery led to further discoveries and to the development of antibiotics.

Directions: Do research to find out more about the life and discoveries of Dr. Alexander Fleming. In the space provided, write a biographical report about Dr. Fleming. If you are unsure about the correct format for a biographical report, check with your teacher. If you need more space, attach another sheet of paper to this page.

Biographical reports will vary. Evaluate the students on the amount of

research they have done and on the effort they put into the report.

_____
_____
_____
_____
_____
_____
_____
_____
_____
_____
_____
_____
_____

Extending Science Concepts

Antibiotics

You read in Chapter 9 that Alexander Fleming discovered the antibiotic properties of *Penicillium* mold. Fleming was experimenting with bacteria. Bacteria are commonly grown in flat, round dishes called *Petri dishes*. The dishes contain the nutrient agar in which bacteria can grow. Agar is a substance similar to gelatin. As the bacteria grow, the microscopic cells reproduce to form colonies, visible on the surface of the agar. Sometimes the bacterial colonies are so numerous that they cover the agar and make it appear cloudy. A clear area on the agar indicates no growth of bacteria. Bacterial colonies growing on nutrient agar are illustrated in Figure A.

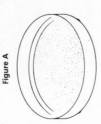

Figure A

Petri dish with bacteria

As Fleming was observing his experiments, he noticed that some Petri dishes had become contaminated by mold. He thought his experiment was ruined and that he would have to throw out the dishes and start over again. Then he looked more closely and made a startling discovery. Bacteria were not growing near the mold. The mold seemed to stop, or inhibit, the growth of bacteria. The agar around the mold was clear—free of bacteria. Today we call this clear area a *zone of inhibition*.

Similar experiments can be used to test the effects of antibiotics on a variety of bacteria. Three molds—X, Y, and Z—are inoculated into 3 Petri dishes, each containing the same type of bacteria. Study Figure B and answer the questions that follow.

(continues)

CHAPTER 9   Protists and Fungi (continued)

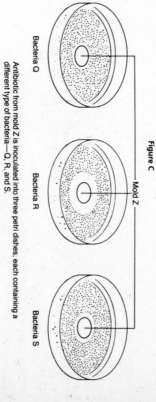

Figure B

Mold X   Mold Y — Bacteria B   Mold Z

Three molds—X, Y, and Z—are inoculated into three petri dishes, each containing the same type of bacteria (B).

1. Which mold seems to have the best antibiotic properties against this kind of bacteria? Explain.

   Z: The zone of inhibition is the largest.

2. Which mold seems to have no antibiotic properties against this kind of bacteria? Explain.

   Y: There is no zone of inhibition.

3. An antibiotic was extracted from mold Z. How should its antibiotic effectiveness be tested?

   Test it against colonies of several types of bacteria, not just the type originally used.

HRW material copyrighted under notice appearing earlier in this work.

47

(continues)

CHAPTER 9   Protists and Fungi (continued)

Antibiotic from mold Z is inoculated into 3 Petri dishes, each containing a different type of bacteria—Q, R, and S.

Figure C

Mold Z

Bacteria Q   Bacteria R   Bacteria S

Antibiotic from mold Z is inoculated into three petri dishes, each containing a different type of bacteria—Q, R, and S.

4. Is antibiotic Z effective against all three types of bacteria? Explain.

   Yes, each dish shows a zone of inhibition.

5. Is the antibiotic equally effective against all three types of bacteria? Explain.

   No, it is most effective against R because this dish shows the largest zone of inhibition.

6. Do you think antibiotic Z would be effective against all types of bacteria? Explain your answer.

   Antibiotic Z would probably not be effective against all types of bacteria. Few antibiotics are. Each type of bacteria has characteristics that make it susceptible or resistant to antibiotics.

48

HRW material copyrighted under notice appearing earlier in this work.

98

CHAPTER 9

Protists and Fungi

Thinking Critically

Directions: Over the last 20 years, the growth of algae in lakes, ponds, and streams has become a significant problem in many areas of the United States. Try to find an example of an algae-overgrowth problem in or near your community. If you cannot find evidence of such a problem in your community, you may choose a recent example of algae overgrowth that has been described in a recent newspaper or magazine. In the space provided, write an essay recommending a permanent solution to the problem. If you need more space, attach another sheet of paper to this page. As you write your essay, be sure to answer the following questions.

• What type of algae is causing the problem?

• What harmful effects are the algae having on the area?

• What is going into the water that causes the overgrowth?

• What efforts have been made by the community to overcome the problem?

• Has any progress been made in solving the problem?

• What other methods could be tried to solve the problem?

The students should identify an area that is experiencing algae overgrowth and suggest possible solutions to the problem.

HRW material copyrighted under notice appearing earlier in this work.

49

CHAPTER TEST 9

Protists and Fungi

Understanding Vocabulary

Write the letter of each organism in the blank next to the classification to which it belongs.

c, e, f, i	1. protozoan	a. *Euglena*
		b. *Penicillium*
		c. ciliate
		d. sporangium
		e. sarcodine
a, g, h, k	2. alga	f. flagellate
		g. kelp
		h. diatom
		i. sporozoan
		j. club
b, d, j, l	3. fungus	k. dinoflagellate
		l. sac

Understanding Concepts

MULTIPLE CHOICE

In the space to the left, write the letter of the choice that best completes the statement or answers the question.

a   4. The means of locomotion used by an amoeba are
     a. pseudopods.   b. cilia.
     c. flagella.   d. vacuoles.

d   5. What is the function of the cell membrane of a protozoan?
     a. movement
     b. capturing food
     c. camouflage
     d. taking in oxygen and releasing carbon dioxide

d   6. Which of the following protozoans are always parasites?
     a. sarcodines   b. ciliates
     c. flagellates   d. sporozoans

(continues)

CHAPTER TEST Protists and Fungi (continued)

b 7. Diatomaceous earth is the empty shells of a type of
   a. protozoan. b. alga.
   c. fungus. d. bacterium.

b 8. Which structure is similar in form to that of a flagella?
   a. tail b. whip
   c. arm d. leg

c 9. Which of these diseases is caused by sporozoans?
   a. amoebic dysentery b. African sleeping sickness
   c. malaria d. influenza

c 10. The mushroom is an example of a
   a. sac fungus. b. imperfect fungus.
   c. club fungus. d. sporangium fungus.

b 11. What characteristic is used to classify algae into different groups?
   a. the ability to perform photosynthesis
   b. the types of pigments contained by the algae
   c. the specific size of the algae
   d. the method of movement by the algae

a 12. Slime molds look like fungi, but they move like
   a. sarcodines. b. algae.
   c. ciliates. d. flagellates.

Interpreting Graphics

In the space, write the letter of the item that best completes each sentence.

a 13. The protozoan shown in this picture is a(n)
   a. amoeba.
   b. ciliate.
   c. flagellate.
   d. sporozoan.

c 14. The *Euglena* shown in this picture moves with the use of its
   a. nucleus.
   b. cell membrane.
   c. flagellum.
   d. chloroplasts.

(continues)

CHAPTER TEST Protists and Fungi (continued)

Reviewing Themes

15. **Systems and Structures**
What adaptations to land life do algae not need?

The students might suggest that algae do not need to be protected from typical temperature and precipitation variations that are experienced by organisms that do not live in water.

16. *Energy*
Are microscopic organisms such as protozoans considered to be part of the food chain? Explain.

Some protozoans feed on various microorganisms and plant materials, and these protozoans are, in turn, eaten by other organisms in the food chain.

Thinking Critically

17. Do you think a sarcodine, a ciliate, or a flagellate would be more likely to lose a race? Why?

The students might suggest that a sarcodine would be more likely to lose a race because pseudopods perform movement less efficiently than cilia or flagella.

18. Why might it be difficult to describe the shape of an amoeba?

The movement of cytoplasm in an amoeba causes the amoeba to change shape almost constantly.

19. Even though algae are plantlike in many ways, algae lack several structures that are found on typical plants. Name three such structures.

Algae do not have stems, roots, and leaves.

20. Explain how algae can be both microscopic and macroscopic.

Although most algae are multicellular and may grow to 50 m or more in length, some algae are single-celled.

21. Why might you find fungi growing on a fallen tree in a forest?

Fungi that are saprophytes feed on dead organisms. Since the fallen tree is dead, fungi are likely to be feeding on it.

22. How are lichens and slime molds similar to fungi?

Lichens are part fungi and part algae. The fungi part of lichens provides support and water retention. Slime molds look similar to fungi, and they reproduce in the same way.

(continues)

Name _____ Class _____ Date _____

Protists and Fungi *(continued)*

Performance Assessment
Observing Fungi

Fungi are important organisms. Just as various animals look quite different from each other, fungi can look quite different from each other and display a wide range of shapes, sizes, and colors.

THE MATERIALS

dry yeast • sugar • methylene blue • 200-mL beaker • medicine dropper • warm tap water • microscope slide • coverslip • compound light microscope

THE INVESTIGATION

Part A. **Readying the materials**

Pour 100 mL of warm tap water into a beaker and add a large "pinch" of sugar. Add dry yeast to the mixture and allow it to stand covered in a warm place for approximately 30 minutes.

The students should correctly follow the procedure.

Part B. **Performing the investigation**

Uncover the mixture and withdraw one or two drops for deposit on a slide. Stain the slide with methylene blue and use a coverslip to cover the slide. Then observe the slide using the different objectives of your microscope.
The students should correctly prepare a slide and mount that slide on their microscopes. Objectives should be carefully switched to prevent slide damage.

Part C. **Gathering your data**

Sketch the fungi you see. On your sketch, label and identify as many fungi structures as you can.

Part D. **Communicating your results**

Compare your sketch with those of your classmates. As a class, suggest reasons why fungi occupy an important niche in Earth's environment.
Help the students to understand that fungi contribute to the ecological balance of Earth.

This task is designed to be completed in one class period. Each group of 3 to 4 students will need the following materials: dry yeast, sugar, methylene blue, 200-mL beaker, medicine dropper, warm tap water, microscope slide, coverslip, and compound light microscope.

53

HRW material copyrighted under notice appearing earlier in this work.

101

UNIT TEST 3

Simple Living Things

Understanding Concepts

MULTIPLE CHOICE

In the space to the left, write the letter of the choice that best completes the statement or answers the question.

a ___ 1. Where are protozoans most likely to be found?
    a. in an ocean
    b. in tropical rain forests
    c. in the atmosphere
    d. underground

c ___ 2. Which algae are the most common single-celled organisms in the oceans?
    a. *Euglenas*
    b. dinoflagellates
    c. diatoms
    d. fire algae

b ___ 3. Choose the word that means the same as toxin.
    a. substance
    b. poison
    c. material
    d. disease

d ___ 4. Which statement about monerans is correct?
    a. Monerans are never helpful to plants and animals.
    b. Monerans are sometimes helpful to plants.
    c. Monerans are sometimes helpful to animals.
    d. Monerans are sometimes helpful to plants and animals.

c ___ 5. Protozoans are classified as
    a. plants.
    b. animals.
    c. protists.
    d. monerans.

d ___ 6. The foods created by algae during photosynthesis provide energy for
    a. the algae.
    b. plants.
    c. animals.
    d. the algae and organisms that eat the algae.

(continues)

UNIT TEST

Simple Living Things *(continued)*

a ___ 7. Where can viruses be found?
    a. Viruses can be found almost everywhere in the world around you.
    b. Viruses can only be found inside the living cells of plants.
    c. Viruses can only be found inside the living cells of animals.
    d. Viruses can only be found inside the living cells of plants and animals.

b ___ 8. Which of these is *not* a locomotive structure of protozoans?
    a. pseudopods
    b. vacuoles
    c. cilia
    d. flagella

d ___ 9. Which of these are responsible for epidemics?
    a. toxins
    b. bacteria
    c. viruses
    d. bacteria and viruses

b ___ 10. Which part of a swimmer's body functions in a similar way to the cilia of a ciliate?
    a. lungs
    b. arms
    c. fingers
    d. toes

a ___ 11. Why are illnesses caused by viruses difficult to treat and cure?
    a. Viruses are not living organisms.
    b. Viruses are difficult to locate.
    c. Viruses are too small to be seen.
    d. Viruses move from one place to another very rapidly.

b ___ 12. In terms of movement, a dinoflagellate is most like which protozoan?
    a. a sarcodine
    b. a flagellate
    c. a ciliate
    d. a sporozoan

c ___ 13. Which of these diseases is caused by viruses?
    a. tetanus
    b. diphtheria
    c. influenza
    d. strep throat

(continues)

UNIT TEST Simple Living Things (continued)

Thinking Critically

19. Name any organism that would not be able to exist without protozoans, and describe how protozoans contribute to the life of that
organism.

**The students might name termites that are not able to digest the wood
they eat without flagellate protozoans present in their digestive
systems.**

20. Some scientists do not agree that algae should be classified as
protists. Is such disagreement healthy for the scientific community?
**Generally, disagreements benefit the scientific community by spurring
additional research.**

21. Viruses can only reproduce inside the cells of living things, but bacteria can reproduce throughout the environment. Do you think a hiker
in the woods would be more likely to acquire a bacterial infection
than a viral infection? Explain.

**Both bacteria and viruses that cause infections are found everywhere
in the world around us so a hiker could contract either one in a forest
environment.**

22. Have viruses ever affected you? If so, how?
**Each student should list one or more viral infections that he or she has
experienced.**

23. Do you think scientists and researchers will ever discover a vaccine
for all known viruses? Why or why not?
**Many students may believe that scientists will be able to develop
vaccines if they are given enough money and time for research.**

24. Why is it said that people who have AIDS die indirectly from the
virus?
**With suppressed immune systems, AIDS victims usually die from
complications of infections such as influenza or pneumonia.**

UNIT TEST Simple Living Things (continued)

___d___ 14. Which of these protozoans cannot move from one place to
another on their own?
  a. sarcodines
  b. ciliates
  c. flagellates
  d. sporozoans

___b___ 15. Which protozoans are most complex?
  a. sarcodines
  b. ciliates
  c. flagellates
  d. sporozoans

Interpreting Graphics

16. Identify each of these protozoans and their locomotive structures.

**A ciliate moves by means of cilia, hairlike structures; a flagellate moves
by means of whiplike flagella; and a sarcodine moves by means of
pseudopods.**

Reviewing Themes

17. *Systems and Structures*
Explain how a lichen functions as a fungus and as an alga.

**A lichen is part fungus and part alga. The fungus part supports the
lichen and helps it retain water, and the alga part produces food for the
lichen through photosynthesis.**

18. *Environmental Interactions*
Explain why a fungus that is a saprophyte is important to a woodland
environment?
**The fungus is an important part of a woodland environment because it
decomposes dead organisms or waste products.**

(continues)

Teacher's Notes

Teacher's Notes

Teacher's Notes